©Copyright 1985

BOOK ORDER INFORMATION
REPAIRMASTER
P.O. BOX 649
WEST JORDAN, UT 84084
BANK CARD ORDERS 1-800-347-5163

HOW TO START AND OPERATE YOUR
OWN APPLIANCE REPAIR BUSINESS.
BOOK ORDER: CALL (800)347-5163
OR ASK FOR FREE INFORMATION
TO BE SENT BY MAIL

EDITORIAL STAFF

Director .... Woody Wooldridge

REPAIR-MASTER® for...
MICROWAVE OVENS and RANGES

VOLUME 1 3050 ISBN 1-56302-000-9

32 VOLUME SET ISBN 1-56302-105-6

PRINTED IN U.S.A.

We have used all possible care to assure the accuracy of the information contained in this book. However, the publisher assumes no liability for any errors, omissions or any defects whatsoever in the diagrams and/or instructions or for any damage or injury resulting from utilization of said diagrams and/or instructions.

FOREWORD

This Repair Master contains information and service procedures to assist the service technician in correcting conditions that are not always obvious.

A thorough knowledge of the functional operation of the many component parts used on appliances is important to the serviceman, if he is to make a proper diagnosis when a malfunction of any part occurs.

We have used many representative illustrations, diagrams and photographs to portray more clearly these various components for a better over-all understanding of their use and operation.

IMPORTANT SAFETY NOTICE

You should be aware that all major appliances are complex electromechanical devices. Master Publication's REPAIR MASTER® Service Publications are intended for use by individuals possessing adequate backgrounds of electronic, electrical and mechanical experience. Any attempt to repair a major appliance may result in personal injury and property damage. Master Publications cannot be responsible for the interpretation of its service publications, nor can it assume any libility in connection with their use.

SAFE SERVICING PRACTICES

To preclude the possibility of resultant personal injury in the form of electrical shock, cuts, abrasions or burns, etc., that can occur spontaneously to the individual while attempting to repair or service the appliance; or may occur at a later time to any individual in the household who may come in contact with the appliance, Safe Servicing Practices must be observed. Also property damage, resulting from fire, flood, etc., can occur immediately or at a later time as a result of attempting to repair or service — unless safe service practices are observed.

The following are examples, but without limitation, of such safe practices:

1. Before servicing, always disconnect the source of electrical power to the appliance by removing the product's electrical plug from the wall receptacle, or by removing the fuse or tripping the circuit breaker to OFF in the branch circuit servicing the product.

NOTE: If a specific diagnostic check requires electrical power to be applied such as for a voltage or amperage measurements, reconnect electrical power only for time required for specific check, and disconnect power immediately thereafter. During any such check, ensure no other conductive parts, panels or yourself come into contact with any exposed current carrying metal parts.

2. Never bypass or interfere with the proper operation of any feature, part, or device engineered into the appliance.

3. If a replacement part is required, use the specified manufacturers part, or an equivalent which will provide comparable performance.

4. Before reconnecting the electrical power service to the appliance — be sure that:

    a. All electrical connections within the appliance are correctly and securely connected.
    b. All electrical harness leads are properly dressed and secured away from sharp edges, high-temperature components such as resistors, heaters, etc., and moving parts.
    c. Any uninsulated current-carrying metal parts are secured and spaced adequately from all non-current carrying metal parts.
    d. All electrical ground, both external and internal to the product are correctly and securely connected.
    e. All water connections are properly tightened.
    f. All panels and covers are properly and securely reassembled.

5. Do not attempt an appliance repair if you have any doubts as to your ability to complete it in a safe and satisfactory manner.

MASTER PUBLICATIONS

TABLE OF CONTENTS

SECTION 1
    Introduction to Microwave Cooking .... 2 thru 6
    Theory and Operation .... 4 thru 6

SECTION 2
    Safety Information .... 7 thru 18
    Measuring Microwave Radiation .... 7 thru 16
    Other Safety Factors 7

SECTION 3
    Westinghouse Microwave Oven .... 19 thru 43
    Diagnosis Charts .... 48 thru 52

SECTION 4
    Early Microwave Ovens .... 44 thru 59
    Diagnosis Charts .... 48 thru 52

SECTION 5
    Amana Radarange .... 60 thru 90
    Diagnosis Charts .... 77 thru 83
    Parts List .... 84 thru 90

SECTION 6
    Frigidaire Microwave Oven .... 91 thru 98
    New Parts 96

SECTION 7
    General Electric Microwave Oven .... 99 thru 118
    Circuit Operation .... 100 thru 103
    Service Procedure .... 103 thru 118

SECTION 8
    Litton Microwave Ovens .... 119 thru 137
    Diagnosis Charts 127
    Service Checks 129
    Schematic 137

FOR IMPORTED MODELS—
ORDER REPAIR-MASTER No. 3053

INCLUDES:

PANASONIC	Private Brands Manufactured by MICRO-TRONICS:
SANYO	
SHARP	Astro Chef
	Imperial
MICRO-TRONICS	Maitre'd
	Mini Chef
	Rich Plan
	Sir Anthony
	United 747

SECTION 1 THEORY AND OPERATION

MICROWAVE COOKING

THEORY & OPERATION
OF THE ELECTRONIC RANGE — SECTION 1

a. The Electron
b. Microwave Energy
c. Conversion of 60 Cycle AC Power to 2,450 Megacycle Power
d. Operation of the Electronic Range

a. THE ELECTRON

The word "electronic" has come into quite common usage during the last few years in describing an electrical device. Just as in the old days a common expression proudly used in describing something new was, "it works by electricity". A salesman in describing to a prospective purchaser that his product works "electronically" builds a sort of mysterious aura around the item in the prospect's mind. We have been exposed to the word "electronic" since World War II and we dimly know it is associated with exciting new things, such as radar and television, for instance. What we probably do not know is that "electronics" has been going on around and within us ever since we were born.

A dictionary defines, "electronics", as "dealing with electrons". Now an electron is a pretty small thing. In fact, it is said to be only a minute particle of negatively charged energy whirling around inside of an atom. Putting the electron to work is what has been done in this new method of cooking, performed by the electronic range.

It is not necessary to learn all there is to know about the electron in order to service the electronic range. It is not necessary to know that the electron is a charge of electricity equal to 4.77 times 10 to the minus 10th power CGS electrostatic units and that its mass is one-one thousand eight hundred and forty-fifth (1/1845) of that of a Proton or hydrogen nucleus. Of course, there is the old saying — "the more you know about a subject, the better you understand it". It is of interest to note that it is a beam of those same electrons that swing across the fluorescent coating of your television picture tube, to give you a picture.

By accepting the existence of the electron, but disregarding the intricacies and detail of "how" and "why" it exists, will greatly simplify our understanding of the operations of the electronic range. All that is required is some basic information to determine if the range is functioning properly.

You are familiar with 60 cycle house current, which requires connecting wires to bring it to appliances, lights, etc. The frequency at which the voltage reverses itself, from positive to negative and back to positive, is 60 cycles per second. At this frequency the electron confines itself mostly to the inside of the wire. If the frequency at which the voltage reverses is increased to, say, 600,000 cycles per second, the electrons become much more active and no longer follow inside the wire. They crowd to the outer surface of the wire and some leave the wire and spread out into space. This is termed radiation and is the principal on which a radio transmitter aerial works. The transmitter generates a frequency high enough so that when power is fed to the aerial, some of the electrons shoot off into space and the result is radio transmission.

These liberated electrons are nothing more than tiny charges of electricity moving at tremendous speeds and, of course, they contain energy. STOP ONE OF THESE ELECTRONS AND THE ENERGY IS TRANSFORMED INTO HEAT. This is what happens in the electronic range, which might be called a radio transmitter, feeding electrons into a "guide" which terminates in the oven compartment. The food to be cooked is placed at the outlet of this "guide". The electrons leaving the end of the "guide" at tremendous speed, strike and penetrate the food. In penetrating the food, they cause molecular agitations and friction in the food which produces heating and cooking.

b. MICROWAVE ENERGY

Energy Waves

Certain types of energy are understood best in terms of their characteristic vibrations. Professionals in the field of foods have little occasion to be concerned with the physics of energy, except with

SECTION 1 THEORY AND OPERATION

MICROWAVE COOKING

the electricity used to power heating devices and with heat itself. In physical terms, ordinary household AC electricity, which vibrates 60 times each second, is said to have a frequency of vibration of 60 cycles per second. Heat is also a form of energy, but by contrast, vibrates at frequencies in the order of 20,000,000 million cycles per second.

Energy waves have measurable lengths, and it is sometimes convenient to describe waves in terms of wave length, rather than frequency. As the frequency of the energy wave increases, the wave length decreases. Just as there are tremendous differences between 60 cycle electrical energy and heat energy, the energy of other frequencies is characterized by physical properties peculiar to particular points in the frequency spectrum.

An inspection of the frequency spectrum will serve as a guide in placing microwave energy in relation to everyday physical phenomena. Above sixty cycles per second in the spectrum is the familiar frequency band of 500 to 1500 kilocycles — (kilo = 1000) — 500,000 to 1,500,000 cycles per second — used for radio reception and marked on the dial of the ordinary home radio. Television uses electromagnetic energy of 50 to 200 megacycles — (mega = million) — 50,000,000 to 200,000,000 cycles per second, and the new UHF (ultra high frequency) stations are on frequencies which run from 470 to 940 megacycles — 470,000,000 to 940,000,000 cycles per second. The length of a typical energy wave at radio broadcast frequencies is 0.3 of a mile long. At UHF TV frequencies, the wave length is about 2 feet.

Visible light is a kind of energy at the high-frequency end of the spectrum with frequencies of the order of 500,000,000 million cycles per second, and wave lengths so tiny that they defy human comprehension. Just below light is infrared heat.

As we view the frequency spectrum, it is important to note that the ability of energy to penetrate various materials varies with frequency. Compared with heat frequencies, lower frequency energy has an improved characteristic in this respect.

Between UHF TV and heat is the region which has been exploited for wartime radar. The electronic range operates in this region at a frequency of 2,450 megacycles — 2,450,000,000 cycles per second — at a wave length of about 5 inches. These wave lengths are so short, when compared to radio waves, that they were called "microwaves" very early in the art. This microwave energy is not heat — it is just energy, hence the cool electronic range and utensils. It has the property of penetrating materials to depths of *about 3 inches,* causing a molecular agitation and friction which produces heating and cooking. The molecules of some materials are more easily set in motion than others. Materials of low moisture content do not respond to this effect in the time required to heat food, hence the ability to cook on china, glass, or paper. Metals act as a reflector of microwaves.

There have been other methods of heating food which are termed "high frequency" or "electronic", which should not be confused with microwave heating. The "induction" method operates on the principle that a coil of wire, energized electrically, sets up an invisible magnetic field or force around it. If food is placed in this alternating magnetic field, some of the magnetic energy will be absorbed by the food and heating will result. Attempts to heat food in this manner have met with very limited success, primarily because of the small cooking space available in any unit of practical design and also because food is not a truly homogeneous mass and tends to accept the magnetic field unevenly, resulting in hot and cold spots. These devices usually employ electrical energy of about 100,000 cycles per second.

Another method is "electrostatic or dielectric" heating. If electrical energy is applied to two metal plates separated by food, the energy from the electrical field established between the plates will be absorbed by the food, producing heat. Devices of this nature again fail from the mechanical standpoint and safety standpoint and they have an inability to heat food evenly. Usually, they have been tried at about 50 megacycles — 50,000,000 cycles per second.

Microwave energy is not hampered by these problems and should be distinguished from "induction" or "electrostatic" methods. The microwave energy is beamed at the food and it penetrates and produces heat. The 2,450 megacycle frequency is

SECTION 1 THEORY AND OPERATION
MICROWAVE COOKING

an excellent compromise between depth of penetration and freedom from uneven heating.

c. CONVERSION OF 60-CYCLE AC POWER TO 2,450 MEGACYCLE POWER

How is the 60-cycle power from a wall fixture converted into 2,450,000,000 cycle microwave power inside the electronic range?

The 115 volt input enters the transformer from the household wall receptacle. The transformer has a two fold purpose. One side of the transformer will step up the voltage, the other side is used as a step down. The secondary voltage is 3.1 volts A.C. and is used to heat the filament of the magnetron. The step up portion of the transformer brings the 115 volt household current up to 4000 volts A.C. The high voltage is connected to the rectifier. The rectifier is a solid state device. Early models of microwave ovens operated on 220 volts A.C. The rectifier was made up of other tubes besides the magnetron and a rectifier filament transformer. The browning unit and browning timer motor were also operated from 220 volts A.C. The rectifier converts the 4000 volts A.C. current into 4000 volts D.C.

During the warm up period, a slightly higher voltage is applied to the magnetron filament, *Figure 2*. Early models which made use of rectifier tubes were supplied with 2.5 volts for the filaments of the power rectifier tubes.

When the electronic timer is in operation, the 115 volt A.C. current is changed to 4000 volts by the transformer. The rectifier converts the high voltage A.C. into high voltage D.C.

Applying this pulsating negative DC voltage to the cathode of a magnetron cooking tube, whose anode is grounded, will cause the tube to generate microwave power at 2,450,000,000 cycles per second. The magnetron tube does not depend on rotating mechanical parts to produce 2,450,000,000 cycle microwave power. This is the second and most important frequency conversion which takes place within the range.

Micro means very small. The wave length of the 2,450,000,000 cycle microwave power is very short — only about 5 inches. This is very short when compared with a typical radio broadcast wave length of three tenths (0.3) of a mile. Hence, the name microwave.

No wires are required to carry microwave power from the magnetron to the food. It can be radiated directly or it can be passed through a "wave guide" into a properly designed oven cavity where the microwave power will be absorbed by food, causing the food to heat very rapidly.

d. OPERATION OF THE ELECTRONIC RANGE

Heating in the electronic range is accomplished by microwave energy at 2,450 megacycles (2,450,000,000 cycles per second) which is generated by a continuous wave, air-cooled magnetron. The microwave energy is directly coupled to the oven cavity, where it is confined by the metal walls and a door designed with appropriate chokes.

Except for the magnetron, all of the equipment in the oven operates on 115 volt, 60 cycle (Hertz) A.C. current. This applies to the browning unit (early models were 220 volts A.C.), power pack and control equipment. On early models the power pack contained four rectifier tubes. The time delay tube used on early models was a heater controlled relay. The rectified voltage is applied directly to the magnetron without filtering, as magnetrons are more stable with this sort of supply than with pure direct current.

Magnetrons are essentially constant voltage devices, like gas voltage regulator tubes, and will draw widely fluctuating amounts of current with very small changes in voltage. Some means of current control is therefore required. Current control is accomplished very simply with a saturable reactor circuit (a current regulator) in series with the primary of the high-voltage transformer. By using this reactor, the magnetron current (and thus the cooking speed) is held virtually constant for changes in line voltage of plus 10% to minus 5% from the design voltage. The reactor is also used to provide lower oven heats by reducing the magnetron current. This method is a particularly convenient way to control heat, as all the switching can be done in the low voltage control circuit rather than in the high voltage power circuit.

SECTION 1 THEORY AND OPERATION

MICROWAVE COOKING

The magnetron and the mercury vapor rectifiers generate a good deal of radio-frequency hash, which, if not suppressed, can cause severe interference with nearby radio and television receivers. Hash filters are built into the range on the magnetron and they reduce the radiated and conducted noise to a point where it is free from objectionable interference. The use of the 2,450 megacycle frequency has been licensed by the Federal Communications Commission for use in the electronic range.

The rest of the circuit is a straight 115-volt control circuit with interlocks to provide protection from high voltage and to stop the radiation if the oven is opened. Air flow is provided for exhausting steam from the oven, cooling the magnetron and other components. Cooling in itself is an important problem and considerable care has been taken to insure long life by running the electrical components at low temperatures. The electronic components in the range operate at relatively high power and must be kept cool if long and trouble-free operation is to be achieved. Cooling air, therefore, is drawn in at the bottom front of the range, where it is most apt to be cool and forced over the electrical components and out the top front of the range and also through the oven door.

MILLIAMPERE TEST ON ELECTRONIC RANGES

A milliampere meter is necessary in setting up an electronic range. The meter shown is a Model 240 Simpson Volt-Ohm Milliamp-meter and can be purchased through radio and TV supply houses. A set of test leads with alligator clips is provided with the meter. These leads are useful in many tests. However, to test the milliampere rate on some electronic ranges, a special set of leads with a phonograph plug must be used. This set can be made up or purchased as an assembly from the supply house.

MILLIAMPERE TEST

(a) The phono-plug is connected to the jack located to the left of fuse. The red lead is connected to the meter jack marked + MA. The black lead is connected to the meter jack marked – MA.

(b) The meter knob is turned to 750 MA (see diagram). Read scale marked DC.

(c) With the oven switch turned to "On" and selector switch on "Hi", *place a china bowl full of water in the oven.* Then turn the minute timer, (on right side), to 2 minutes. The meter should read between 310 and 320 milliamperes (see diagram).

(d) The milliamp rate can be adjusted by a variable resistor located under the bright cap fastener near the fuses. Turn to the left to raise and to the right to lower.

CAUTION: *DO NOT MOVE THE METER KNOB WHEN THE METER IS CONNECTED TO MILLI-AMP LOAD.*
DO NOT HANDLE THE PHONOGRAPH PLUG OR THE METER WHILE THE OVEN IS HEATING.

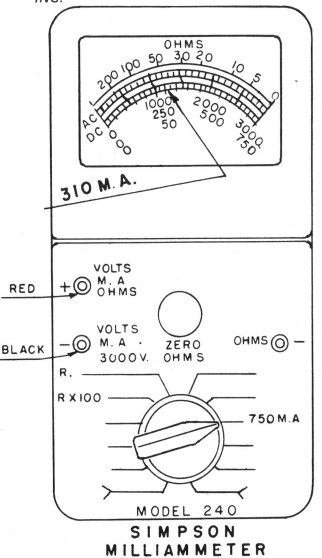

SIMPSON MILLIAMMETER

SECTION 1 THEORY AND OPERATION

MICROWAVE COOKING

**IMPORTANT
SAFETY INFORMATION**

The basic design of radar and microwave ovens and ranges makes it inherently safe to use as well as to service. However, there are some necessary precautions that must be followed in servicing these units.

1. The unit MUST be adequately grounded. A three prong outlet box will suffice, providing the outlet box is grounded properly. Have the outlet checked by an electrician prior to use. Do not defeat the purpose of a grounded outlet by the use of an adapter.
2. Check the door seal before using. Look for physical damage within the choke area before energizing the unit. A dollar bill test, as used on a refrigerator door gasket, will work equally well on a radar or microwave oven.
3. The unit should be disconnected when removing side or back panels. Follow instructions closely when making "live power tests", as indicated in this Repair-Master.
4. Discharge all capacitors. Use insulated wire or tools for this purpose, as shown in illustration, *Figure 23*. The shock may not hurt but the reflex action could compound the injury.
5. When testing "live" have the utmost respect for the HIGH VOLTAGE area in and about the magnetron tube. Keep the area clear of anything that would induce an arc or ground.
6. Do not defeat the purpose of the door or interlock switches or mechanisms; these components can be tested and checked with the unit disconnected. When replacing these switches they must be adjusted to disconnect at a distance not to exceed 1/2 inch upon opening of the door. This is measured at the top of the door.
7. Always use good safety practice, account for all of the tools that were used before replacing panels.

CAUTION: *When cooking in the Microwave oven it is advisable that the container being used is greater in diameter than the depth. If the diameter is smaller than the depth, the surface tension of the liquid will restrict the boiling for a time. The container and its contents will still be heating however. The internal pressure will overcome the surface tension quite suddenly and the liquid will blow out of the container. This is especially true when heating soups or broth. When making the efficiency test of the oven, be sure the container used for water is wide and flat. The liquid will not damage the oven but there may be clean up work to do.*

SECTION 2 SAFETY INFORMATION

MEASURING MICROWAVE RADIATION

SERVICE INFORMATION

OPERATING INSTRUCTIONS FOR THE NARDA 8100 RADIATION MONITOR

The purpose of the monitor is to check the radiation leakage around the microwave oven door or other places where radiation could possibly occur.

The instrument measures radiation leakage in milliwatts per square centimeters (MW/CM^2). A water load of 275 cc., approximately 1-1/3 cups water, is to be placed in the oven and used as a load during leakage test.

Operate the instrument on its internal rechargeable battery or a 115 volt power supply. It may also be charged from a 115 volt, 60 hertz power source. Use the instrument on the 2450 MHZ switch position. The "Meter Response" switch should be set on fast position. A slow setting requires too long a time to register. The "Alarm" control should be set on 50 which sounds an audible alarm when the meter reads 50% of full scale deflection. This is to provide warning against high levels which may damage the instrument.

The "Range" switch may be used on the lower or higher meter scale. On an oven with an unknown leakage, use the high scale first. Switch to the low scale for low leakage. The "Test" switch is used to check the battery and the probe. If either is faulty, the meter needle will not read above the "Test Minimum" mark on the meter. The zero control is used to zero the needle.

With the probe and cone spacer plugged into the instrument, turn the monitor on-off switch to the on position. Check battery and probe "Test" switch. If battery reading does not come up to "Test Minimum" setting, plug in the AC cord. If probe test fails, do not use probe.

During usage, if the probe becomes inoperative or disconnected, the audio alarm will come on.

The test probe *MUST* be held by the grip portion of the handle, otherwise, a false reading may result if the operator's hand is between the handle and the probe.

Hold the probe perpendicular to the cabinet door. Place cone of the probe on the door and/or cabinet-door seam and move along the seam. If the leakage of the oven is unknown, move the probe slowly. If at any time the audible alarm sounds, proceed with care in order to exceed a full scale reading of the meter or remove the probe from the area of the leakage. When testing near a corner or access area of the door, keep the probe perpendicular to the areas, making sure the probe end at the base of the cone does not get closer than 2" to any metal. If it does, an erroneous reading may result.

Always use the 2" spacer with the probe. Also, always proceed carefully in areas of high leakage or the probe will be accidently burned-out. The rotating stirrer causes high peaks of energy. Although the meter has averaging capabilities, the probe will react instantaneously to peak power changes which will cause the burn-out.

If the oven is likely to have a large amount of leakage, approach the oven slowly with the probe, while observing the meter. This is achieved by holding the probe 2 to 3 feet from the oven and then moving toward the oven surface or gap between the door and oven body while observing the meter. When high leakage is expected, do not move the probe horizontally along the oven surface, this could cause possible probe burn-out. The greatest leakage is generally found at the corners. After the maximum leakage is established to be within the meter scale range, then the probe may be moved horizontally around the door surface.

When the probe is stored in its carrying case, cover the end of the probe with aluminum foil to prevent accidental burn-out of the probe from stray fields of radiation.

MEASUREMENT, METER, and TECHNIQUES FOR THE MICROWAVE OVEN

To properly test microwave radiation emitted from a microwave oven, the correct measurements and techniques must be employed to obtain accurate readings.

The Narda Surveyor — Model 8100 is an approved instrument for measuring microwave radiation from microwave ovens. It is manufactured by the

SECTION 2 SAFETY INFORMATION

MEASURING MICROWAVE RADIATION

Narda Microwave Oven Corporation, Plainview, Long Island, New York. This instrument will measure only microwave frequencies and not other radiation energies such as X-rays, light rays, heat rays, etc. Other instruments are required to measure these frequencies.

THE INSTRUMENT

The Narda Surveyor — Model 8100

The instrument consists of three basic components: the probe, the spacer, and the meter and power supply cord.

The Probe

The meter is available with three probes for measuring the power density of microwave radiation.

Blue Probe	0 to 0.2 mW/cm^2 and 0 to 2 mW/cm^2
White Probe	0 to 2 mW/cm^2 and 0 to 20 mW/cm^2
Red Probe	0 to 20 mW/cm^2 and 0 to 200 mW/cm^2

You will note in the chart above that each probe has two ranges of measurement. The range selector switch is located on the meter itself and will determine the reading on the mW/cm^2 scale.

The probe must be plugged into the meter before any reading can be obtained. This connection is made below the range switch and can be inserted only one way. A handle is provided on the probe for holding the probe without interference in the reading.

The Spacer

The probe is calibrated with the polyethylene foam cone spacer in place. Therefore, it must be used when making measurements. When the Narda 8100 instrument is ordered, it comes equipped with the correct cone spacer for measurements at 2450 MegaHertz. For measuring 915 MegaHertz a separate cone must be used. This cone is similar to the 2450 spacer, but contains two crossed antennas within its face for proper calibration at 915 MegaHertz. The 915 MegaHertz spacer should never be used at 2450 MegaHertz or an erroneous high meter reading will result.

The Meter

The meter is a delicate instrument and should be handled with care. The meter may be operated on either the supplied rechargeable battery or on 115 volt A.C. The meter will always be operating on the rechargeable battery unless it is plugged into 115 volt A.C. To do this, simply loosen the thumbscrew on the top of the meter, slide the cover out, and plug in the power supply cord.

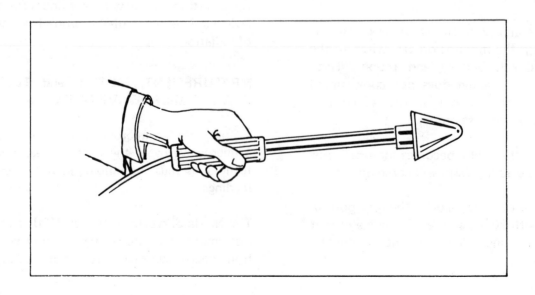

SECTION 2 SAFETY INFORMATION

MEASURING MICROWAVE RADIATION

You will also note there is a switch for 115 volt or 220 volt. This switch should be placed in the 115 volt position. The battery will always be recharged whenever the meter is plugged into 115 volt A.C.

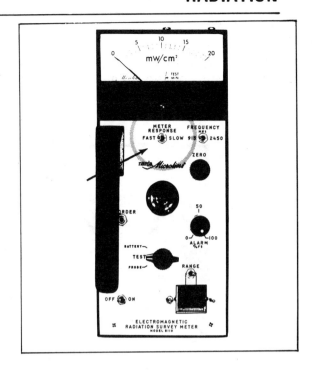

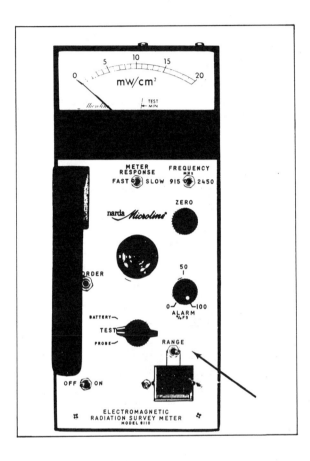

FREQUENCY

The frequency switch should be placed on 2450 for measurement of most charts, microwave ovens, read spec.

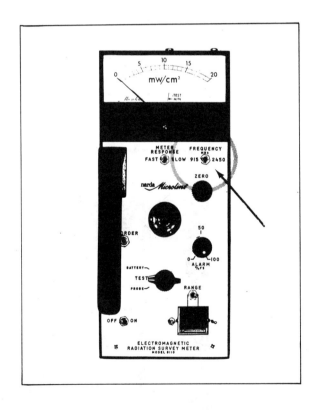

METER Response

The meter response switch should be set on "fast". This allows a faster response of the meter. If the meter response switch is set on "slow", the probe must be moved very slowly and in many cases a several second delay will be required for the meter to register maximum radiation.

SECTION 2 SAFETY INFORMATION

MEASURING MICROWAVE RADIATION

ON-OFF Switch

To operate the instrument, the on-off switch must be in the "on" position. This should be turned "off" after testing is completed.

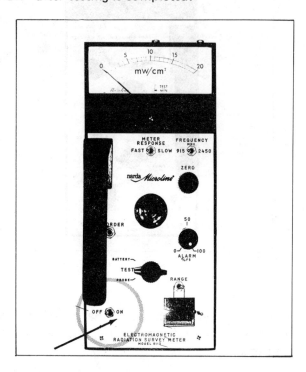

ALARM

The audible alarm is set at any percentage of full scale value. This alarm will buzz when this percent is indicated by the meter. For field survey work, the alarm level should be set at 100%.

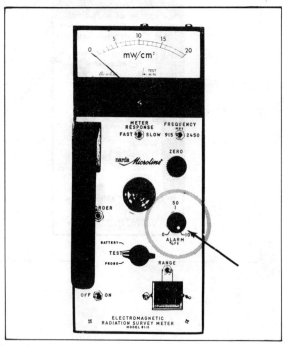

ZERO Adjust Knob

Because the instrument is extremely sensitive it is necessary to "zero" the meter before attempting to make any measurements by setting the needle to "0" with the control knob. At times the meter may drift from the zero position. This may occur when the battery requires recharging. It may also occur just after the battery has been fully charged. In this case the meter should be turned on for about 15 minutes after which it will hold steady at zero. The zeroing function has to be performed with the microwave oven off. The zero should be checked after making a measurement to make sure the instrument did not drift.

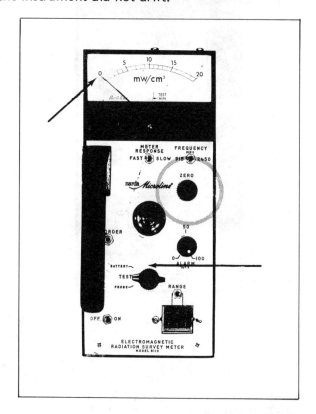

TEST Knob

To test the instrument for proper operation, the test knob should be turned to the "battery" position. If the needle goes to the right of the "test minimum line" on the meter, this indicates there is enough power in the battery to properly operate the instrument. The probe can also be tested in the same manner by turning the test knob to the "probe" position.

SECTION 2 SAFETY INFORMATION

MEASURING MICROWAVE RADIATION

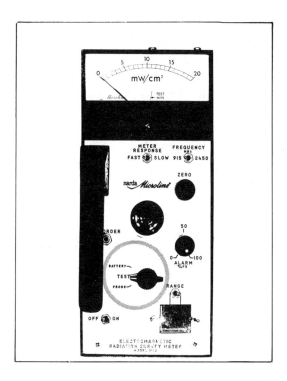

RANGE Selector and Reading

When using the white probe, you will be using the 0 to 20 scale or 0 to 2 scale. The preliminary readings should be made on the 0 to 20 scale. If the preliminary readings are less than 2 mW/cm², you can then change to the 0 to 2 scale which, of course, will give you a more accurate reading. For help in reading the scale, refer to the following chart.

Care must be taken to read the meter scale correctly. Using the white probe and the range switch positioned on the 0-2 scale, the divisions represent 1/10 of a mW/cm² or .1 mW/cm². If the needle is half way between 0 and .1, then the reading is .05 mW/cm². If the needle is half way between .1 and .2 then the reading is .15 mW/cm². On the 0-20 scale the small divisions represent 1 mW/cm². If the needle is half way between 0 and 1, then the reading is .5 mW/cm². If the needle is half way between 1 and 2, then the reading is 1.5 mW/cm², etc.

TESTING FOR MICROWAVE RADIATION

A standard procedure has been developed for measuring radiation from microwave ovens. This procedure has been approved by both industry and government, all of whom are interested in proper microwave oven radiation measurements. This procedure is detailed below and must be followed to obtain accurate measurements.

1. *Place 1-1/3 cups (275cc) of water in the center of the oven cavity.* The container should be of glass, paper, or foam. The use of this water provides a small load while making measurements. Even though it is possible to obtain a

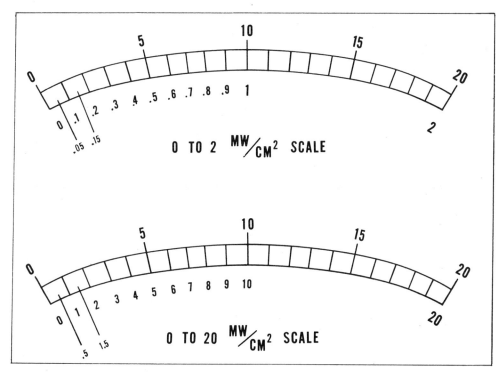

SECTION 2 SAFETY INFORMATION

MEASURING MICROWAVE RADIATION

slight increase in radiation from an oven without a load or with the load not placed in the center, the proposed government standards are based on 1-1/3 cups (275cc) of water placed in the center of the oven. This standard must be followed. Do not operate the oven without a load; the oven will overheat and the life of the magnetron will be shortened.

If the water boils and the oven fills with steam, a lower reading will result on the meter. Never allow the water to boil. If this occurs, refill the container with cold water.

2. *The sensor of microwave energy which is in the end of the probe should be 2 inches from the oven for each measurement.* This is accomplished by use of the spacer cone provided. This sensor, when positioned properly, will accurately detect and measure microwave emission. When used incorrectly, erroneous measurements will result. The sensor must not be positioned closer than 2 inches from any surface. This includes oven handles and style contours. The minimum clearance between the oven and adjacent walls, cabinets, and other appliances is 20 inches.

3. *The probe must be positioned perpendicular to the metal surface of the oven for proper measurements.* (See diagrams on following page.)

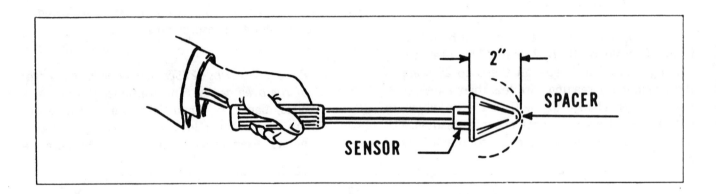

SECTION 2 SAFETY INFORMATION

MEASURING MICROWAVE RADIATION

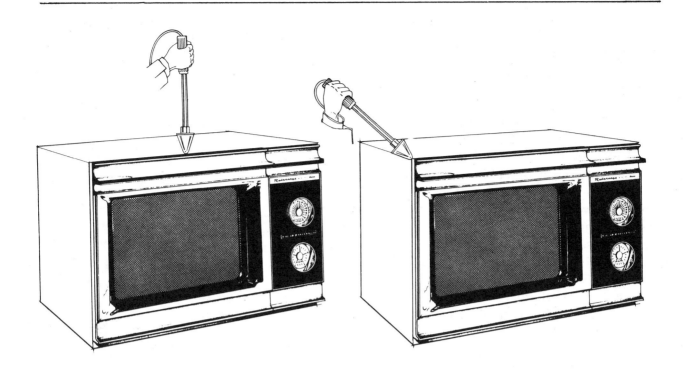

13

SECTION 2 SAFETY INFORMATION

MEASURING MICROWAVE RADIATION

For proper operation and testing, the following procedure should be followed:

1. Plug probe connector into meter.
2. Turn on meter.
3. Adjust ZERO knob so needle reads zero.
4. Turn ALARM switch to 100.
5. Turn TEST switch to BATTERY. Needle should move to right of TEST MIN. mark. If not, operate on 115 volt A.C.
6. Turn TEST switch to PROBE. Needle should move to right of TEST MIN. mark. If not, meter is inoperable.
7. Turn METER RESPONSEE switch to FAST.
8. Turn FREQUENCY switch to 2450.
9. Turn RANGE switch to 2 or 20 as read on probe connector. Readjust ZERO knob so needle reads zero.
10. Move oven away from side walls and overhead cabinets. Clearance required around the probe is 6 inches. It would be preferable to place oven on open table.

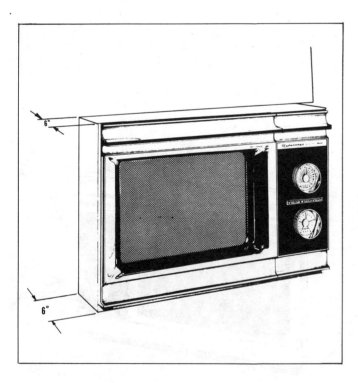

11. Always use the 2 inch cone spacer for measurements at 2450 megaHertz. For 915 MegaHertz cone must be changed.
12. Always use 1-1/3 cups (275cc) of water placed in the center of the oven when measuring.
13. Hold probe by handle only.
14. Hold probe perpendicular to oven at all times before reading meter.
15. In screen portion of door hold probe at a 45° angle to corner. Measurements in inside corners should be made with care.

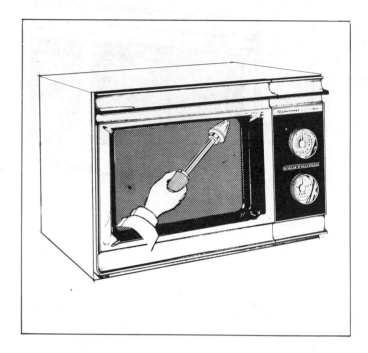

CAUTION — Erroneous measurements may occur.

1. Do not attempt to make a measurement in a confined space. An oven with cabinets above the door and/or a wall adjacent on the left should be moved to an open table before measurements are attempted. It may only be possible to pull the oven out and away from the wall and cabinets. This is satisfactory.
2. Do not hold probe at an angle to metal surface of oven.
3. Do not place probe closer than 2 inches from any oven projection such as handles or style contours.
4. Do not wedge probe between oven and wall.

SECTION 2 SAFETY INFORMATION

MEASURING MICROWAVE RADIATION

IF YOU HAVE A PROBLEM

PROBLEM	CAUSE or ACTION
1. Meter is turned on and audio alarm sounds when not making a measurement.	Caused when probe connector is not inserted properly or probe defective.
2. Battery TEST, meter needle to left of TEST MIN.	Battery needs recharging.
3. Probe TEST, meter needle to left of TEST MIN.	Probe defective.
4. Meter needle drifts from zero even after resetting with zero knob.	Battery needs recharging.
5. Meter needle drifts from zero after battery recharged.	Leave meter on for 15 minutes to stabilize voltage.
6. Alarm sounds during battery or probe test.	Set alarm to 100%.

TECHNIQUE

Measurements on the microwave oven should be made in two ways:

1. With oven door normally closed
2. With oven door open to the interlocks

Place a 1-1/3 cup (275cc) water load in the center of the oven. In order that all doors are closed in the same way, open the door fully. Then, holding the door handle, close to within 2 inches of the fully closed position and release the door. The door springs will carry the door to a closed position. If all doors are closed in the same manner, better comparison may be made between ovens. Hold the probe by the handle. Make certain the probe cable and the 115 volt A.C. cord are not near the area being measured. Move the probe, attached to the 2 inch spacer, perpendicular to the oven.

To determine the maximum door opening before the interlocks turn the oven off, shims are used. These are made of 1/16" plastic, 1" wide by 2-1/2" long. Ten shims are placed between the door and the oven and the start button pushed. Shims are then removed one at a time until the oven operates. A count of the shims gives the distance the door will stay open and oven operate. Avoid inserting the shims into the vinyl gasket area between door and oven.

In measurements with the door open, again use 1-1/3 cups (275cc) of water. Care must be taken not to insert the probe spacer into the gap between the door and oven. If this is done, there will no longer be a 1 inch spacing between the oven and the end of the probe. This will result in an erroneous measurement.

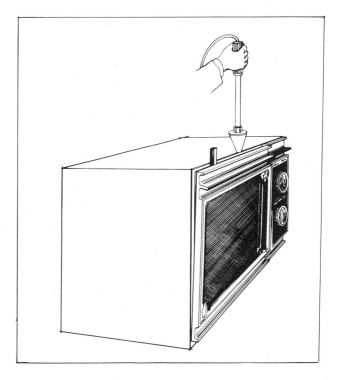

If all of the above procedures are followed, accurate measurements will result.

SECTION 2 SAFETY INFORMATION

MEASURING MICROWAVE RADIATION

DATA SHEET

Data sheets should be used for recording radiation levels. Blanks are provided for recording meter readings as the probe is moved around the perimeter of the door. The probe is to be moved into the approximate areas of the door as represented by the blank. A search is then made in this area for a maximum reading, which is then recorded.

Although a drawing of the entire oven is shown on the data sheet with arrows leading to blanks for recording readings, the arrows are to be interpreted as if they were radiating from the top, bottom, and sides of the door.

Measurements are made with the "door in the normally closed position" and also at "maximum door opening before unit shuts off".

Record at the top of the data sheet "maximum radiation door closed" and "maximum radiation door open".

A diagram and blank is provided for recording "maximum door opening before unit shuts off".

A check on the interlocks can be made by ear. The interlock switches click as they open and close. Also note proper start — stop switch operation. Record "OK" in the blanks if operating.

Circle the Narda meter "range" and "time constant". Time constant refers to "meter response" switch. Also circle the oven "frequency".

Fill in the other blanks on the data sheet. Always sign and date the data sheet.

By recording the results of tests made, a complete and accurate record of units can be obtained for permanent use.

SECTION 2 SAFETY INFORMATION

MEASURING MICROWAVE RADIATION

Form A760-1

Max. Radiation Door Closed __1.0__
Max. Radiation Door Open __NONE__
Length of Service __1 YEAR__
Apparent Cause of Leakage Any Over 5mw/cm² __NONE__

¼ Maximum door opening before unit shuts off.

Reading at maximum door opening before unit shuts off

Reading door in normal closed position

Readings around oven (values shown on diagram):
Top corners: .5, 0, 0, 0
Top edges: 0, 0, 0, 0
Sides upper: .4, .1, 0, 0
Sides middle: .7, .2, 0, 0
Sides lower: .4, .3, 0, 0
Bottom: .3, .2, .2, 0
Bottom edges: .7, .8, .9, 0

SPECIMEN

Amana Radarange® MICROWAVE OVEN

DATA SHEET

Customer Name: __John Doe__
Address: __35 Lovell__
__Chicago, Ill.__
Model No. __RR-1__
Serial No. __R 1835000__
Estimated Operating Time/Day __10__ Minutes
Interlock Test: Start-Stop __OK__
  Left __OK__
  Right __OK__
Survey made with NARDA 8110 with white probe. Test load 275 cc of water.
Range: (0-20) or 0-2
Time Constant: (fast) or slow
Frequency: (2450) or 915
(Circle settings used.)

Past Service _____

Reason Unit was Tested _____

__P. Houston__ __1-5-69__
Signature of Surveyor   Date

Amana Refrigeration, Inc., Amana, Iowa, Subsidiary of Raytheon Company

17

SECTION 2 SAFETY INFORMATION

MEASURING MICROWAVE RADIATION

Form A760-1

Max. Radiation Door Closed _____
Max. Radiation Door Open _____
Length of Service _____
Apparent Cause of Leakage Any Over 5mw/cm² _____

Maximum door opening before unit shuts off.

Reading at maximum door opening before unit shuts off

Reading door in normal closed position

Amana Radarange® MICROWAVE OVEN

DATA SHEET

Customer Name: _____
Address: _____
Model No. _____
Serial No. _____
Estimated Operating Time/Day _____ Minutes
Interlock Test: Start-Stop _____
Left _____
Right _____
Survey made with NARDA 8110 with white probe. Test load 275 cc of water. (Circle settings used.)
Range: 0-20 or 0-2
Time Constant: fast or slow
Frequency: 2450 or 915

Past Service _____

Reason Unit was Tested _____

Service Company _____
Address _____

Signature of Surveyor _____ Date _____

Amana Refrigeration, Inc., Amana, Iowa, Subsidiary of Raytheon Company

SECTION 3 WESTINGHOUSE MICROWAVE OVEN

SERVICE PROCEDURE COMPONENT DATA

WESTINGHOUSE MICROWAVE OVEN OVEN POWER PERFORMANCE TEST

To test the oven for power performance, a Pyrex cup or dish that will hold at least two cups of water and a thermometer with a range over 180° will be needed.

1. Pour at least two cups of water into a pyrex vessel.
2. Record the temperature of the water.
3. Place the container and water in the center of the glass tray in the oven.
4. Close the door and latch.
5. Set the timer from 2½ to 3 minutes.
6. Push the start button, with a stop watch or accurate time piece allow two minutes to elapse, then shut oven off.
7. Remove container and test temperature rise of the water using the thermometer.
8. If the oven is operating properly at line voltage (115 V.) there should be approximately a 55° temperature rise.

The voltage at which the oven is operating will greatly effect the results of this test. It would be well to check the voltage prior to the test.

OPERATIONS AND TEST
CONTROL CIRCUIT, Figure 1

Connect oven to power source prior to tracking circuit.

1. With door open the oven light will be on. The light circuit is from one side of line through the left interlock switch. It then continues on to the oven light and then to the other side of the line.
2. Upon closing the door, both the left and right interlock switches are closed. The light should be off at this time. The light can be turned on, if necessary, by pushing the oven light switch to the closed position. When the oven light switch is in this position, a circuit is completed from one side of the line through the oven light switch. It continues on to the oven light and then to the other side of the line.
3. With the door closed, by pushing down on the latch lock switch, it will close both the start switch and the interlock switch. By turning on the timer, a circuit will be completed to the primary winding of the transformer. Now, follow the circuit from one side of the line through the left interlock and the latch interlock switches. The circuit continues through the start switch, the cavity thermal cut-out, and the relay and tube thermal cut-out, then to the primary winding of the transformer. It follows across the other side of the start switch and then to the other side of the line.
4. With the door closed and the timer on, there is also a circuit through the timer motor and to the dial light. A circuit is also completed to the fan motor and the relay.

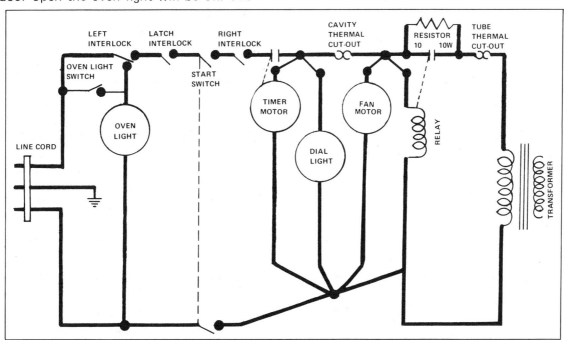

Figure 1 — Control Circuit

SECTION 3 WESTINGHOUSE MICROWAVE OVEN

SERVICE PROCEDURE COMPONENT DATA

THE SECONDARY CIRCUIT *(High Voltage)*

CHECKING THE RECTIFIER

See trouble shooting chart page 35.

Included in the circuit is a by-pass resistor to protect the service technician against a high voltage shock while taking milliamp readings. DO NOT REMOVE THIS RESISTOR NOR OTHERWISE DEFEAT ITS PURPOSE, except when checking the internal impedance of the meter. It is only necessary that this be done once for each meter being used to take readings. To find the impedance of the meter being used, it will be necessary to take a reading with the resistor both in and out of the circuit. The test follows:

1. The unit must be shut off and unplugged.
2. Remove the end panel.
3. Remove black ground wire from the right hand side of the resistor. see *Figure 2*
4. Connect positive meter lead to the right side spade of the resistor and the negative motor lead to the loose end of the black ground lead. THE GROUND CONNECTION SHOULD NOT BE REMOVED AT THE BULKHEAD.
5. Adjust meter to proper scale setting.
6. Place a cup of water in the oven.
7. Connect power and start oven.
8. Make note of the meter reading and shut unit off.
9. Disconnect power.
10. Remove the 10 ohm, and 10 watt resistor from the circuit.
11. Reconnect power, start oven, repeat test and make note of reading.

EXAMPLE OF METER READING

Using a Simpson Model 260

1. Milliamp reading with resistor in the circuit — 275 MA.
2. Milliamp reading with resistor removed from circuit — 300 MA.
3. The difference between the two readings shows a meter impedance of 25 MA.
4. When using this meter to adjust the milliagrams of a unit, set the adjustment at 275 MA to compensate for the meter impedance.

TESTING THE MAGNETRON

Magnetron failures have been generally identified and are grouped into various categories as shown on left side of *Figure 3*.

INSULATION BREAKDOWN

1A. Usually found at the end of the leads. A breakdown will take place when the insulation or the material deteriorates or degrades. This occurrence is very rare and takes place where very high temperatures prevail for extended periods.
1B. Will also be found in the region of the bead insulation. It may occur if the beads rest against the metal filter box. When a tube is installed, care should be exercised to prevent the beads from resting against the metal walls.

AIR *Figure 3*

This happens when the glass envelope of the tube is cracked or destroyed and instead of a vacuum, air has now entered the envelope. The result will be internal arcing and high line current if high voltage is applied. Noticeably missing will be the black, shiny getter flash, *Figure 3*. The getter will be white or entirely removed or invisible. If failure of the tube took place when in use, the copper antenna will lack the characteristic shine but oxidation will take place, resulting in a dull copper color. There are several causes of this failure.

1. *Getter Crack* — is the result of a poor getter to glass spacing. The vacuum envelope will rupture on the side of the cathode glass, directly opposite the tip of the getter rod.
2. *The Tipoff* — or a tipoff crack could result if the tube were struck due to careless handling. It will sometime, but very seldom, break off or crack from thermal or mechanical stress. The crack may be around the tipoff or in length as shown in drawing.
3. *Suck-In* — is the result of excessive high power on the glass. Atmospheric pressure pushes the glass inward toward the vacuum until a hole is formed. Using the range improperly, such as heating an empty oven or using metal cooking utensils, is generally to blame for this fault. The damage will generally be found at the

SECTION 3 WESTINGHOUSE MICROWAVE OVEN

SERVICE PROCEDURE COMPONENT DATA

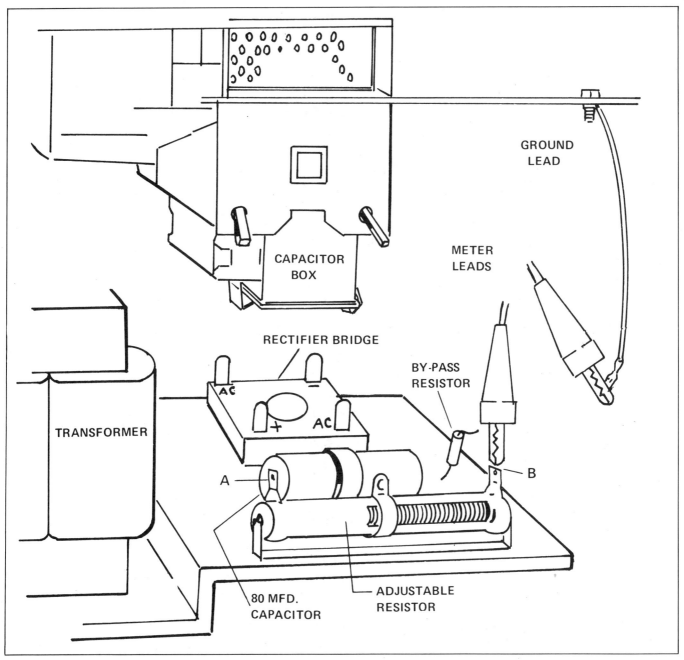

Figure 2 — Connecting the Meter

CAUTION: *The by-pass resistor has been added to the unit circuit for your protection while taking meter readings or making adjustments to the unit.*
DO NOT UNDER ANY CONDITIONS TEST OR ADJUST THE UNIT WITH THE BY-PASS RESISTOR DISCONNECTED OR REMOVED FROM THE CIRCUIT.

SECTION 3 WESTINGHOUSE MICROWAVE OVEN

SERVICE PROCEDURE COMPONENT DATA

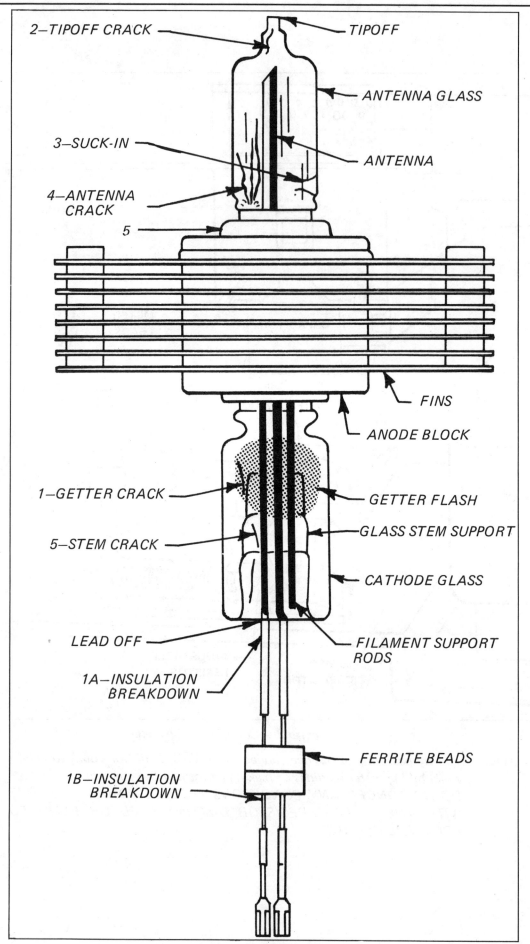

Figure 3 — The Magnetron

SECTION 3 WESTINGHOUSE MICROWAVE OVEN

SERVICE PROCEDURE COMPONENT DATA

antenna glass-to-metal seal.

4. *Antenna Crack* — is the result of mechanical stress. Cracks can be found lengthwise in the glass and there may be several. The metal ridge will be shiny and the anode block will be depressed.

5. *Stem Crack* — will result under thermal stress. The crack will be lengthwise between either two of the filament support rods or it may be across the rods slightly above the pressed glass.

OPERATIONAL FAULTS

1. *Open Heater* — can be found with the use of an ohmmeter after the transformer leads are disconnected. Resistance is normally less than one OHM; the filament will not be necessarily shorted internally. When a tube is removed for any reason, it should be checked, since usage will cause the filament to become fragile and handling of the tube could break the filament.

2. *Low Emission* — is a direct cause of "wearout" of the emission characteristics of the directly heate cathode. It can be detected by delayed tube current. It may take a much longer time to get to the operating point of 300 MA than when the tube was new. Also, the tube current will not get high enough to cause oscillation of the tube with normal line voltage. The current will be somewhere in the range of 200 MA to 250 MA and the oven produces low power into a load, of times two thirds or less than normal.

PARTS REPLACEMENT AND ADJUSTING

Following the text and outline below will make the job of adjusting or replacing parts on the Westinghouse Microwave Oven much easier. The first part of this text will deal with the services that can be performed from the front of the oven. DISCONNECT POWER TO OVEN.

Timer Knob, Dial Light and Oven Light

1. Remove knob by pulling forward.
2. Remove the Phillips screw and pull the right side of the bezel (trim) toward you. Move bezel to the right to clear the outer case.
3. Remove the retainer plate to replace the oven light push button or the outer door latch.
4. Remove screw to replace the escutcheon clip.
5. Push in and turn to remove dial lamp (bayonet type base).
6. Reach into oven light recess, push and turn to remove oven light.

Grease Shield Removal

1. Disconnect power, open oven door.
2. Place screw driver into slots in front edge of shield and pull back to release pins, see *Figure 4*.
3. Remove shield from oven.

Stirrer Removal, *Figure 5*

1. Disconnect power, remove grease shield from oven (see Grease Shield Removal text).
2. Remove the LEFT HAND THREADED SCREW from shaft.
3. Lift the stirrer out. In replacing, be careful that the stirrer fan is not inverted. The paddles must be down.

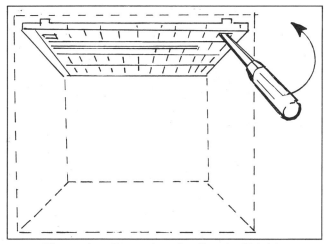

Figure 4 – Removing the Grease Shield

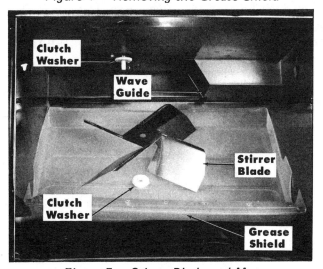

Figure 5 – Stirrer Blade and Motor

SECTION 3 WESTINGHOUSE MICROWAVE OVEN

SERVICE PROCEDURE COMPONENT DATA

CAUTION
There are some precautions which must be followed in servicing microwave ovens to maintain safety.

1. The unit must be operated from an adequately grounded outlet. *Do not use a two wire extension cord, nor defeat the purpose of a grounded plug by removing the ground prong.*
2. The door gasket and choke area must be examined for physical damage before energizing the oven. DO NOT operate the oven until any damage is repaired.
3. Always disconnect the power when removing panels. When making a "live" test, take heed of the warnings within this manual. Do not reach into the equipment area while the unit is energized.
4. Check and tighten all connections prior to energizing the oven.
5. Always discharge capacitors against the filter box with a screw driver before working in the high voltage area, *Figure 23(A).*
6. Always have the filter box in place, except when it must be removed to replace the tube or other components.
7. DO NOT DEFEAT THE PURPOSE OF THE INTERLOCK SWITCHES *at any time.*
8. DO NOT OPERATE THE OVEN WHEN IT IS EMPTY.

ACCESS PANEL
The access panel is at the rear of the unit and is secured by screws and rivets. One of the rivets is covered with an aluminum seal. The rivets can be drilled out. To gain access to the internal parts of the unit, the access panel must be removed. This panel must be replaced using a pop rivet gun. DO NOT REMOVE THIS PANEL UNLESS YOU HAVE THOROUGHLY READ THIS REPAIR MASTER.

BLOWER MOTOR AND SCROLL, *Figure 6*

Disconnect power to oven.
1. Disconnect the two motor leads.
2. Remove the motor bracket screws that secure the bracket to the oven.
3. Slide the motor and scroll assembly to the rear of the oven and remove it from the cabinet.
4. Remove the blower housing and screws and lift out.
5. Remove blower from shaft using 1/8 Allen set screw wrench.
6. Remove motor mount screws and separate motor from bracket.
7. When reinstalling, the hub of the blower wheel must be flush with the end of the motor shaft for proper clearance and operation. The motor fan is press fitted to the motor shaft.
8. The lip of the scroll must fit under the lip of the oven cavity body.

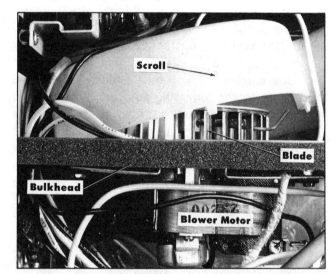

Figure 6 — Blower Motor and Scroll

SECTION 3 WESTINGHOUSE MICROWAVE OVEN

SERVICE PROCEDURE COMPONENT DATA

TRANSFORMER AND COMPONENTS, *Figure 7*

Listed below are the parts that make up the transformer assembly.

Transformer
The transformer is used both as a step-up and a step-down conversion of power as follows:
A. Step-up converts 115 V.A.C. to 3100 V.A.C.
B. Step-down converts 115 V.A.C. to 3 V.A.C.

Filter Capacitor, *Figure 8*
The filter capacitor is used to by-pass the AC current around the magnet coil, this allows a wave-free D.C. current to flow in the magnet coil, resulting in a nearly wave-free constant magnetic field.

Bridge Rectifier, *Figure 8*
The bridge rectifier will convert the 3100 V.A.C., supplied by the transformer, to 4000 peak volts DC. This voltage is then applied between the cathode and the anode of the magnetron tube during operation.

Adjustment Resistor, By-pass Current, *Figure 8*
To adjust the scope of the magnetic field that is supplied to the magnetron, an adjustment resistor is used. Adjusting the resistor will regulate the current in the magnetic coil, which in turn effects the magnetic field. This, in turn, also effects the power of the magnetron. The resistor is covered to protect it against accidental shorting or by-pass. To adjust beyond that provided by the 7500 ohms resistance would cause improper performance of the tube, resulting in faulty operation and possible damage to the unit.

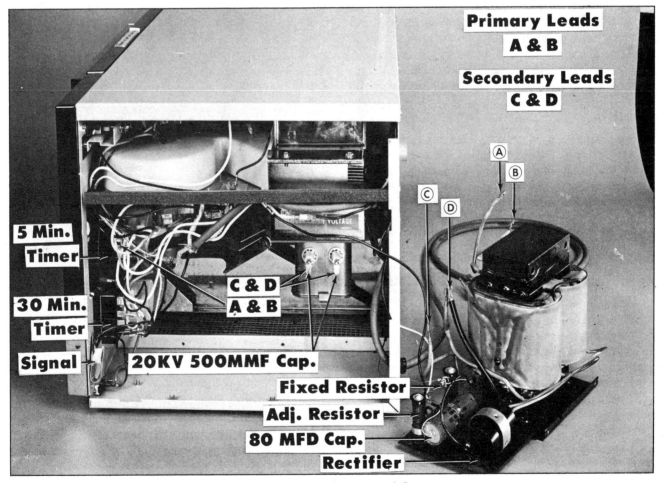

Figure 7 – Transformer and Components

SECTION 3 WESTINGHOUSE MICROWAVE OVEN

SERVICE PROCEDURE COMPONENT DATA

Figure 8 — Magnetron, Transformer and Components

1. Fan Cover
2. Fan Motor
3. 5 Minute Timer
4. 30 Minute Timer
5. Signal
6. Signal "off-on" switch
7. Transformer
8. Wave Guide
9. Magnetron Tube
10. R/F Filter Capacitor
11. Filter Box Assembly
12. Bridge Rectifier
13. Fixed Resistor-2.5KΩ
14. Adj. Resistor-5KΩ
15. Filter Capacitor-80MFD-450V

SECTION 3 WESTINGHOUSE MICROWAVE OVEN

SERVICE PROCEDURE COMPONENT DATA

By-Pass Resistor, *Figure 2*

A by-pass resistor is incorporated in the high voltage circuit to protect the service man when taking meter readings or adjusting the unit.

When installing a new transformer, all of the above must be removed and reinstalled with the new transformer. These individual parts should be tested before using. The parts can be replaced without changing the transformer.

REPLACING COMPONENTS

Bridge Rectifier, *Figures 1 and 9*
1. Remove the leads, remove the center screw.
2. Lift unit out and replace, as shown in *Figure 2*.
3. Double check wiring for correct hook-up.

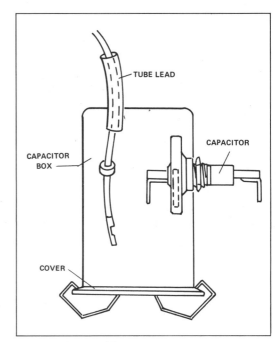

Figure 10 — Capacitor Check

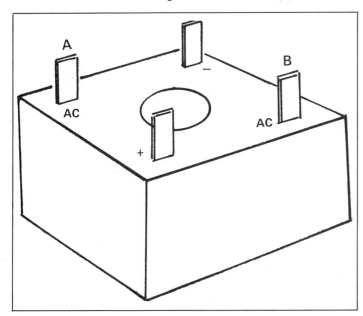

Figure 9 — Checking the Rectifier

Filter Capacitor, *Figure 10*
1. Using a soldering iron, disconnect the leads to the resistor.
2. Lift resistor out of bracket.
3. Rivets must be removed to replace the bracket.

By-Pass Resistor
1. Using a soldering iron, remove the lead that is soldered to the resistor.
2. Remove the screw that holds the other wire and the resistor to the frame.

Transformer, *Figure 7*

Discharge the capacitor with a screw driver to the filter box, *Figure 23(A)*.
1. Remove the screws and nuts that secure the transformer to the base pan.

2. Remove the high voltage cover from the capacitor and remove the leads from the capacitor. Capacitor should be discharged.
3. Remove the other connections, lift out transformer frame before reassembly.
4. Double check wiring for proper hook-up.

MAGNETRON THERMAL CUT-OUT, *Figure 11*
1. Remove leads from the transformer and relay.
2. Slide cut-out and clip from the tube fin.
Replace in reverse order, double check wiring.

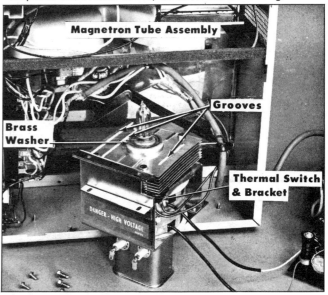

Figure 11 — Removing the Magnetron

27

SECTION 3 WESTINGHOUSE MICROWAVE OVEN

SERVICE PROCEDURE COMPONENT DATA

R.F. CAPACITORS

A small amount of the R.F. current may feed back down the tube filament leads. To prevent this from happening, the ferrite rings and by-pass capacitor is included in the circuit. If this were allowed to feed back to the line voltage, it would result in radio and T.V. interference within close proximity. Any of the R.F. current that escapes the ferrite rings is conducted through the capacitor to the inside walls of the junction.

Capacitor Replacement, *Figure 12*

Discharge capacitor as previously outlined.
1. Remove the high voltage cover.
2. Remove clip and cover from the can bottom.
3. Support the capacitor and disconnect the leads.
4. Remove the mounting nut and remove capacitor from the can.
5. Reassemble in reverse order, adjust the grounding clip to secure the capacitor tightly.

MAGNETRON, FILTER BOX AND COIL HOUSING, *Figure 13*

The magnetron tube converts DC power to RF power and requires a high DC voltage at 4000 volts and a low AC voltage of 3.1 volts within a magnetic field for normal operation. Incorporated in the tube is an anode and a cathode, which is also the filament.

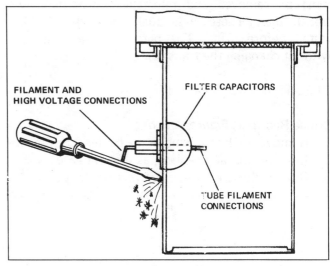

Figure 12 – Discharging the Capacitor

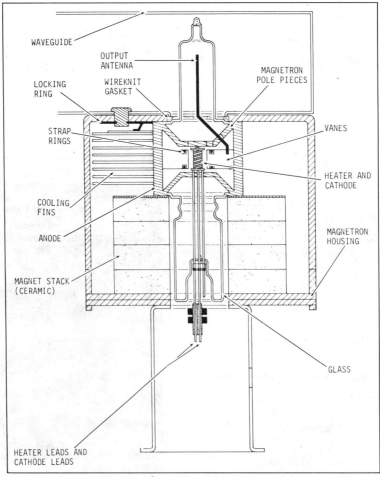

Figure 13 – Magnetron Assembly

SECTION 3 WESTINGHOUSE MICROWAVE OVEN

SERVICE PROCEDURE COMPONENT DATA

For proper operation, electrons must move from the cathode to the anode, *Figure 14*. A high voltage must exist between the cathode and the anode to cause the electrons to flow. The cathode must be heated. Heating is accomplished with the use of a filament. The filament is supplied with 3 volts that come from the step-down portion of the transformer. As a diode, this must be converted to a magnetron. This is accomplished by a magnetic field applied parallel to the cathode. The magnetic field causes a rotary action of the electrons as they spin about the cathode, rather than traveling in a straight line from cathode to anode.

This spinning action of the electrons and the configuration of the anode is responsible for R.F. currents to flow on the surface of the anode. An antenna connected to the anode conducts the R.F. power down the wave guide.

To replace
1. Disconnect coil, ground the thermal cut-out leads.
2. Remove the transformer and its components.
3. Support the tube assembly, rotate the torsion bars ¼ turn in either direction and remove from housing.
4. Remove tube, filter box and coil assembly using caution to protect the tube from breakage.

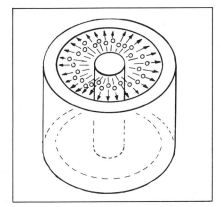

Figure 14 — Heater and Cathode

REPLACING TUBE

1. The tube, brass washers and housing, along with the capacitor box, should be assembled into one unit.
2. Properly position the brass washers on the tube.

3. Place assembly into magnetic housing on the oven.
4. While holding assembly in place, insert both tension bars, the short bends should be toward the oven cavity. Lock the bars by turning ¼ turn, *Figure 15*.

The tube should be handled as gently as a television tube. The utmost care should be taken in removing and replacement. Hold the tube by the fin area only to avoid damage.

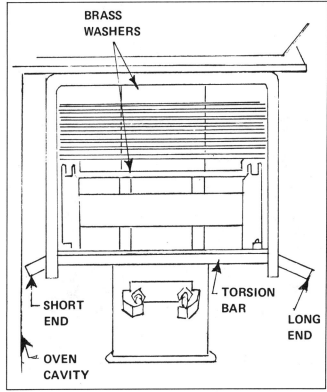

Figure 15 — Positioning Brass Washers

RELAY, *Figure 54*

1. Remove the leads.
2. Loosen right hand screw, remove left hand screw and lift relay from unit. Install new relay in reverse order.

REMOVING OVEN CAVITY

1. Remove access panel.
2. Remove two screws securing the oven cavity to the base frame.
3. One screw can be found on the right side of the transformer, and the second screw is to the right of the rectifier.

SECTION 3 WESTINGHOUSE MICROWAVE OVEN

SERVICE PROCEDURE COMPONENT DATA

4. The oven cavity can now be pulled forward approximately 3 inches without disconnecting the wiring.
5. Some components can now be tested.

INTERLOCK SWITCH TESTING, *Figure 16*

1. Check with an ohmmeter. The interlock switch is normally open. With the door open, the light should be on.
2. Place a cup of water in the oven. Close the door and push the latch down. Turn the timer knob to energize timer.
3. Timer will advance if it is operating correctly. If the interlock is inoperative, the unit will not function.

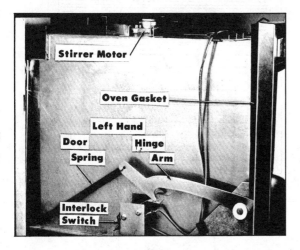

Figure 16 — Interlock Switch Location

RIGHT HAND INTERLOCK

1. Place a cup of water in the oven.
2. Close the door and energize timer.
3. Grasp the door handle and pull out on the handle to de-energize the switch.
4. If the switch is functioning properly, the unit will stop operating. It would be well to check all interlocks after servicing the unit.

RIGHT HAND INTERLOCK REPLACEMENT

1. Disconnect the switch leads.
2. Remove the nuts securing switch bracket to bezel.
3. Remove the screws holding the bracket to the switch.

OVEN DOOR AND COMPONENTS

1. Releasing the spring from the door shackle assembly will allow the shackle to be lifted over the door stop lever.
2. Remove the screws and washers securing the top hinge plate to bezel.
3. Remove the two bolts and nuts and washers that secure the bottom hinge plate to the weldment, see *Figure 17*.

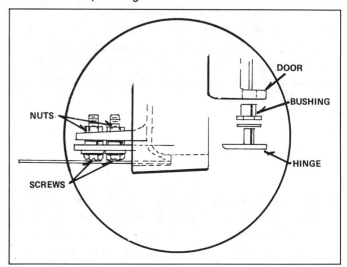

Figure 17

4. Slide door assembly away from bezel while holding bottom hinge plate corner of the door. Thread the door stop lever through the opening in the bezel.
5. Remove hinge plates and components from door assembly.
6. Remove black door gasket from door assembly, *Figure 18*.
7. Remove the two Phillips screws securing the handle to the door.
8. Using a 7/64 Allen wrench, remove the socket head screws.
9. Separate the inner and outer door assembly.
10. Remove the door handle and strike.

When reassembling the door, it is recommended that a clear RTV cement be used along the top edge of the door. Also, 3 M #465 adhesive transfer tape should be used to replace the gasket around the door.

SECTION 3 WESTINGHOUSE MICROWAVE OVEN

SERVICE PROCEDURE COMPONENT DATA

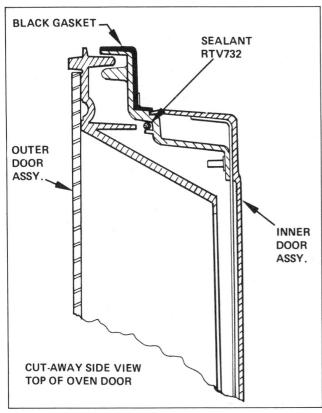

Figure 18 — Cut-Away Viewing Top of Oven Door

NOTE: Some models use roll pins to secure the inner and outer door assemblies, Figure 19. Drive the pins into the door to separate these parts.

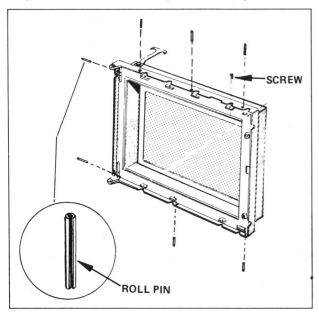

Figure 19 — Roll Pin Location

To replace or test the following, it will be necessary to remove the capacitors and the power cord. The capacitors should be discharged as previously outlined. Other leads that are in the way must be disconnected and the components removed or pushed to one side.

TIMER Figure 24(E)

1. There are two screws securing the timer to the bezel, they must be removed. This also frees the dial light bracket.
2. Remove the wire leads, one lead at a time, and connect to new timer.
3. Be sure light bracket is properly aligned when reassembling.

AIR BAFFLE

If the air baffle is lifted at the rear and moved to the rear, it can be lifted out.

LATCH INTERLOCK SWITCH, Figure 24(L)

1. Remove air baffle.
2. Remove two screws and wire leads, lift out.

LATCH SWITCH, Figure 20

1. Remove air baffle.
2. Remove the four screws that secure the latch switch bracket to the bezel.
3. Remove switch through front of bracket.
4. Remove wiring, connect wiring to new switch.
5. In reassembling, double check the component parts for proper installation such as the lock bar and spring, inner door latch, and retainer. Then assemble the latch switch and bracket with the door latch over switch arm.

OVEN LIGHT SWITCH, Figure 21

1. Remove air baffle.
2. Remove leads from connectors.
3. Remove securing nut from switch.
4. Check for proper clearance of the toggle switch.

SECTION 3 WESTINGHOUSE MICROWAVE OVEN

SERVICE PROCEDURE COMPONENT DATA

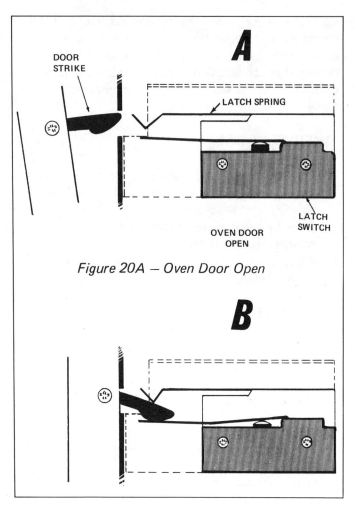

Figure 20A — Oven Door Open

Figure 20B — Oven Door Closed

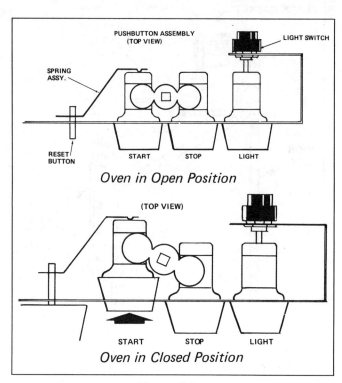

Oven in Open Position

Oven in Closed Position

Figure 21

SECTION 3 WESTINGHOUSE MICROWAVE OVEN

SERVICE PROCEDURE COMPONENT DATA

OVEN LIGHT SHIELD AND SOCKET

1. Remove air baffle.
2. Remove aluminum foil strips and clips.
3. Disconnect light socket leads.
4. Assembly can now be removed. Be sure aluminum foil strips are used when reassembling.

THERMAL CUT OUT

1. Remove air baffle.
2. Remove oven light shield and socket.
3. Remove clip and lift out.

LEFT INTERLOCK SWITCH, Figure 22

1. Remove air baffle.
2. Remove the mounting screws from switch.
3. Remove the wire leads, assemble to replacement switch.

CAUTION: *The by-pass resistor has been included in the unit circuit as a safety measure. DO NOT DEFEAT ITS PURPOSE BY REMOVAL.*

RECTIFIER BRIDGE CHECK

1. Place volt ohmeter on a setting to read ohms. Use a x 10,000 scale or higher.
2. Check resistance between the A and plus, and the A and minus. Reverse the leads and check once more. The difference should be a minimum of one half of the scale between the two readings. If there is no change in reading when the leads are reversed, replace the bridge.
3. The bridge can be checked without removal from the unit, however, the leads must be disconnected.

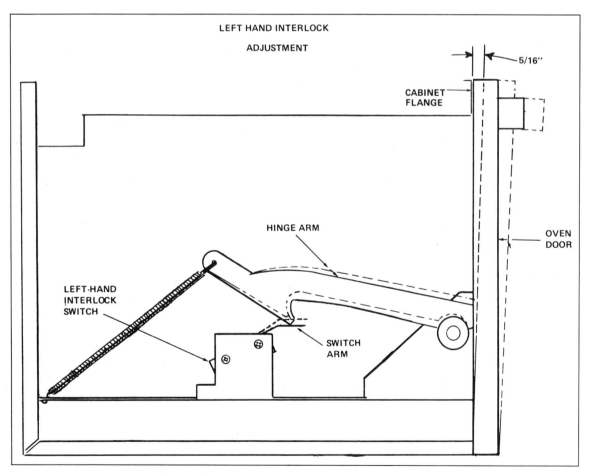

Figure 22 — Adjustment of Left Interlock'

SECTION 3 WESTINGHOUSE MICROWAVE OVEN

SERVICE PROCEDURE COMPONENT DATA

CHECK R.F. CAPACITOR FAILURE AS FOLLOWS, *Figure 23.*

1. Un-plug oven.
2. Remove access service panel. (Three sheet metal screws hold service panel to underside of base pan.)
3. Discharge R.F. capacitors per *Figure A.* (There may be a cover over the R.F. capacitors; spread clips to remove cover.)
4. Disconnect transformer and high voltage leads from R.F. capacitor terminals per *Figure B.*
5. Remove filter box cover and disconnect magnetron tube filament leads from the R.F. capacitors.
6. Check R.F. capacitors per *Figure C.*.
7. Remove mounting nut and washer from defective R.F. capacitor and remove the capacitor from within filter box assembly per *Figure D.*
8. NOTE: *When replacing the new R.F. capacitors, it is very important that you handle and hold the capacitor by its expanded or blue section because of the possibility of damaging the insulation between the plates.*

Take care in placing the capacitor through its mounting hole located on the filter box assembly so as not to damage the wiring terminals. Connect the magnetron tube filament leads to the R.F. capacitor.

In making the outside electrical connections to the R.F. capacitor, refer to *Figure B.*

9. Re-install service panel.

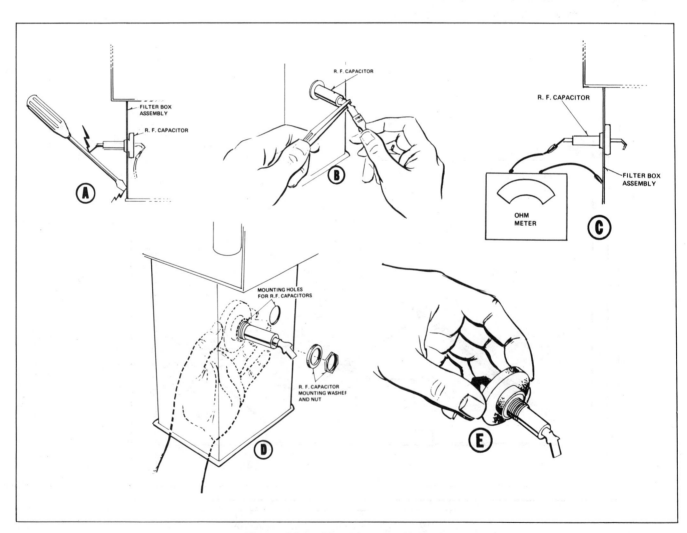

Figure 23 — Checking the R.F. Capacitors

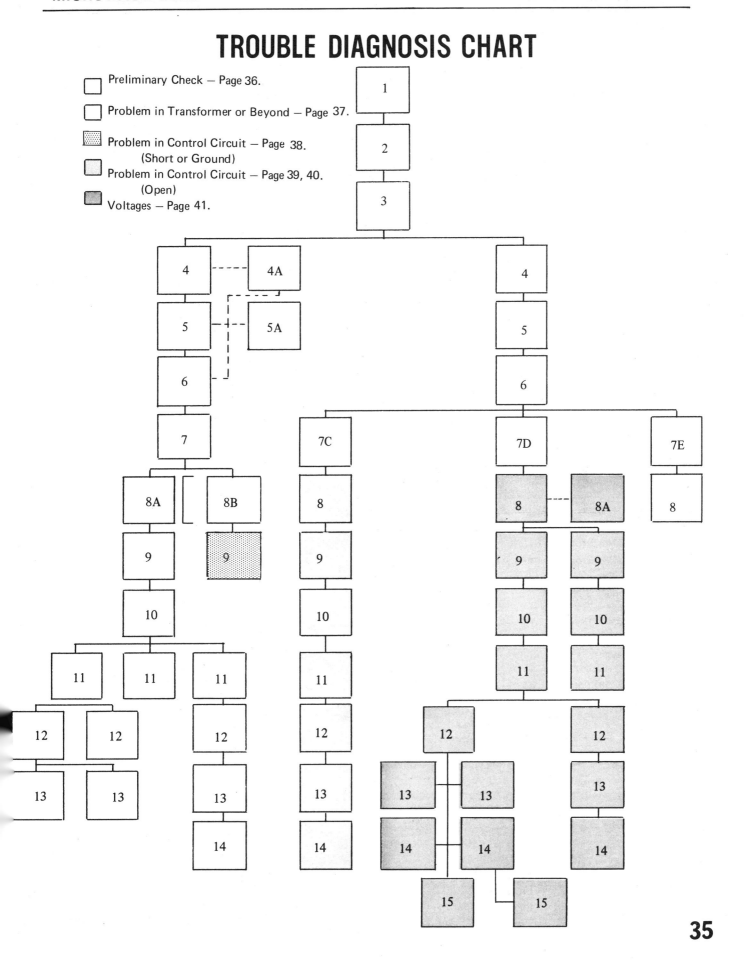

SECTION 3 WESTINGHOUSE MICROWAVE OVEN

SERVICE PROCEDURE COMPONENT DATA

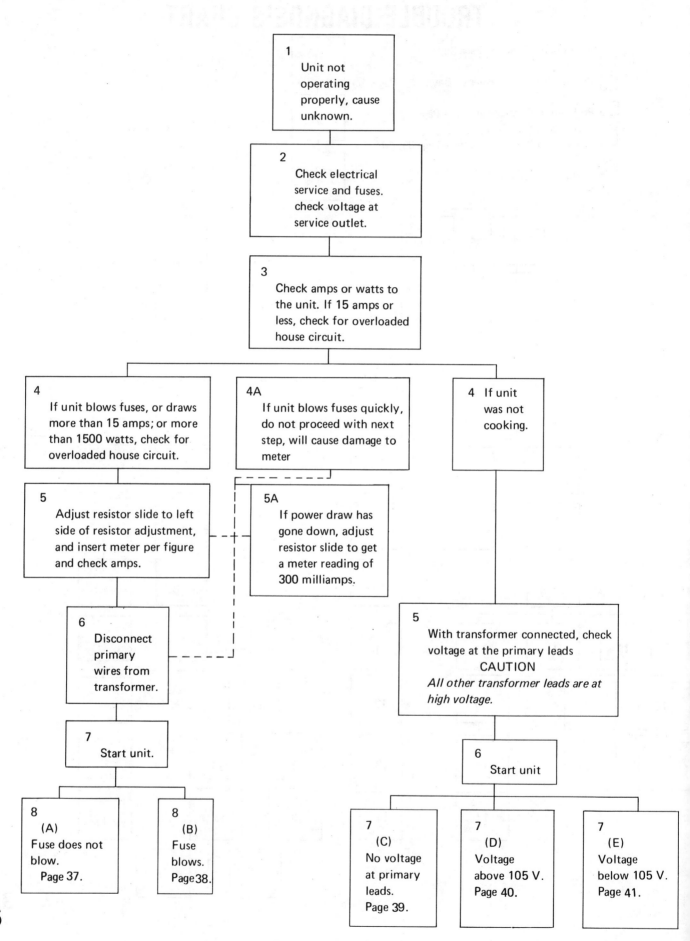

SECTION 3 WESTINGHOUSE MICROWAVE OVEN

SERVICE PROCEDURE COMPONENT DATA

THIS SECTION PERTAINS TO PROBLEMS IN THE TRANSFORMER OR BEYOND.

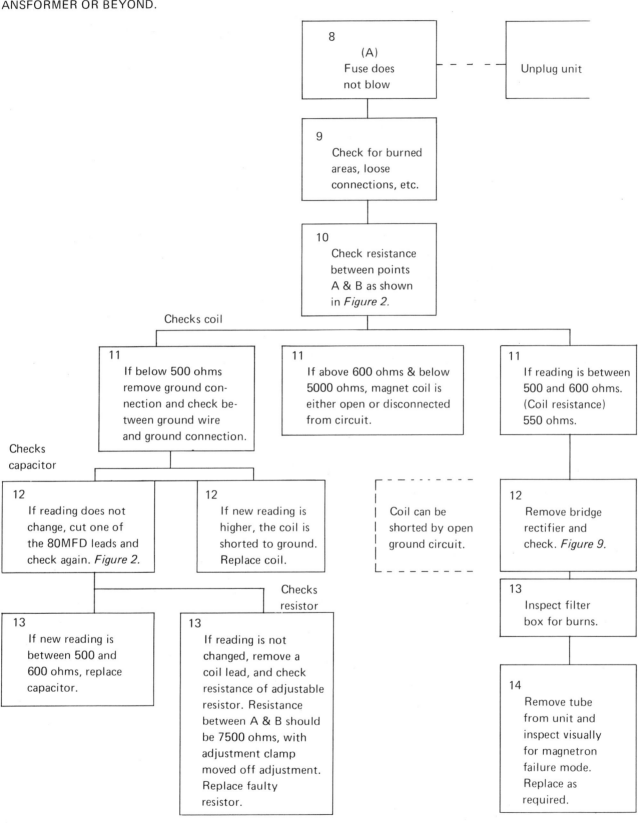

SECTION 3 WESTINGHOUSE MICROWAVE OVEN

SERVICE PROCEDURE COMPONENT DATA

THIS SECTION PERTAINS TO CHECKING CONTROL CIRCUIT FOR SHORTS AND GROUNDS.

8 (B) Fuse blows

9 Check for loose connection or grounded wires. The timer and switches by themselves draw less than 1 AMP. If the unit blows a fuse while the transformer is disconnected, there will be some burned areas visible. Clean and tighten connection.

SECTION 3 WESTINGHOUSE MICROWAVE OVEN

SERVICE PROCEDURE COMPONENT DATA

THIS SECTION PERTAINS TO CHECKING FOR OPEN CIRCUIT.

7 (C) No voltage at the transformer primary leads

8 With unit *not* plugged in. Close door, set timer, place start switch in the lock-ccok position.

9 Check for continuity tube thermal switch lead at transformer splice to white (gray) power lead connection. No continuity, thermal switch or relay defective.

10 Check continuity across cavity thermal switch. No continuity, switch open or defective. Requires minimum of 20 minutes to reset.

11 Check continuity of right interlock. No continuity, switch defective.

12 Check continuity across start switch. No continuity, switch defective.

13 Check continuity across start switch interlock. No continuity, switch defective.

14 Check continuity across left interlock. No continuity, switch defective.

SECTION 3 WESTINGHOUSE MICROWAVE OVEN

SERVICE PROCEDURE COMPONENT DATA

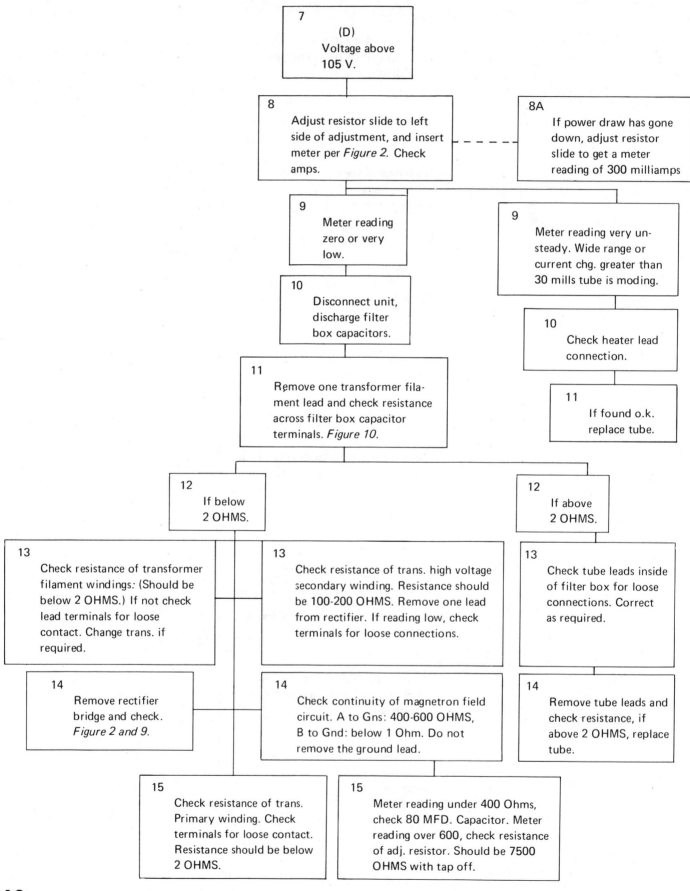

SECTION 3 WESTINGHOUSE MICROWAVE OVEN

SERVICE PROCEDURE COMPONENT DATA

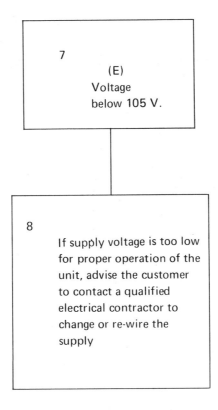

SECTION 3 WESTINGHOUSE MICROWAVE OVEN

SERVICE PROCEDURE COMPONENT DATA

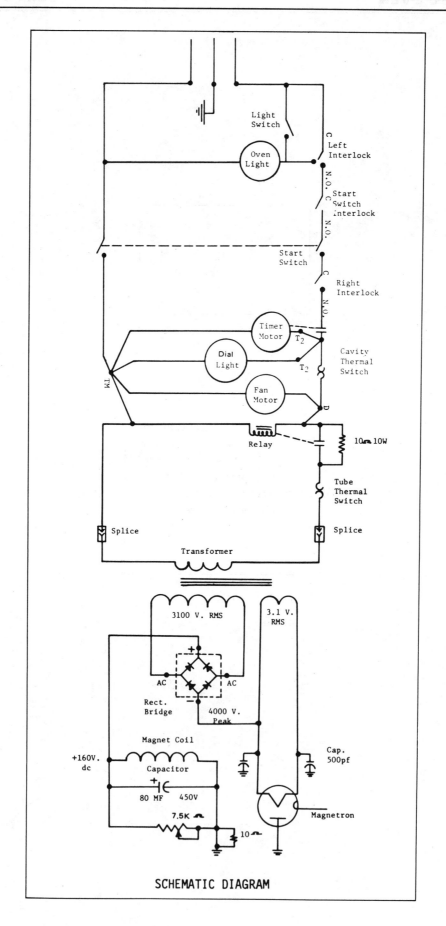

SCHEMATIC DIAGRAM

SECTION 3 WESTINGHOUSE MICROWAVE OVEN

SERVICE PROCEDURE COMPONENT DATA

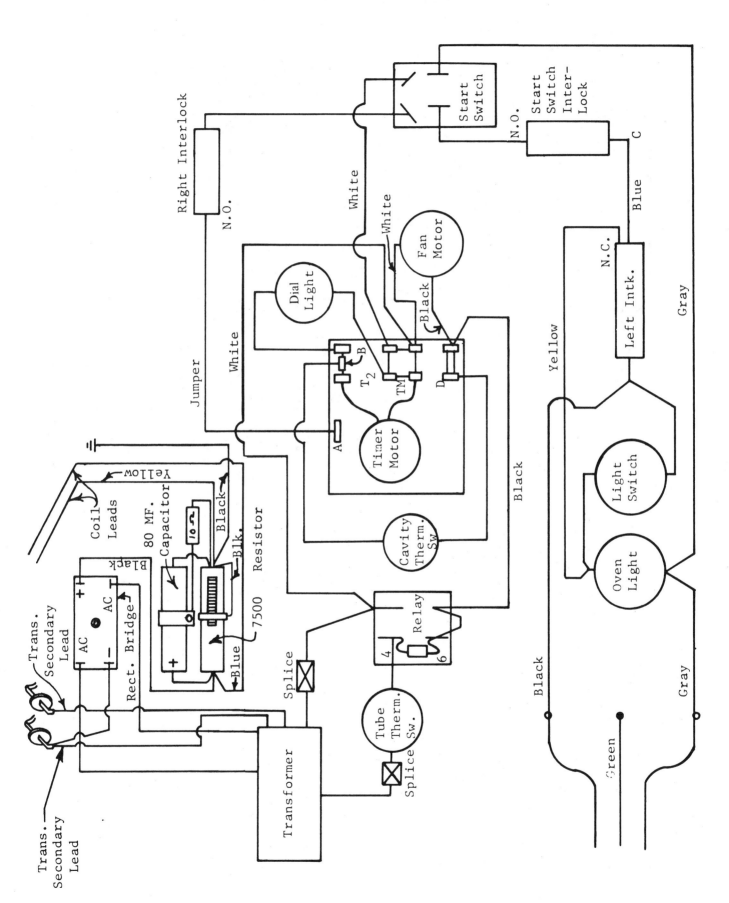

SCHEMATIC DIAGRAM

SECTION 4 EARLY MICROWAVE OVENS

SERVICE PROCEDURE COMPONENT DATA

CONNECTION DIAGRAM

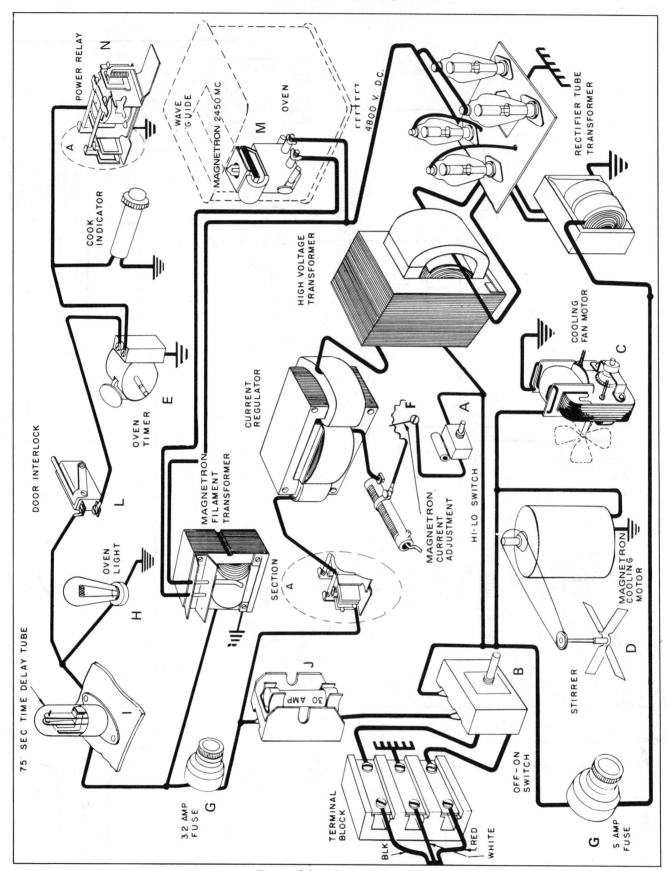

Figure 24 – Connection Diagram

SECTION 4 EARLY MICROWAVE OVENS

SERVICE PROCEDURE COMPONENT DATA

POWER SUPPLY CHASSIS AND MAGNETRON PARTS

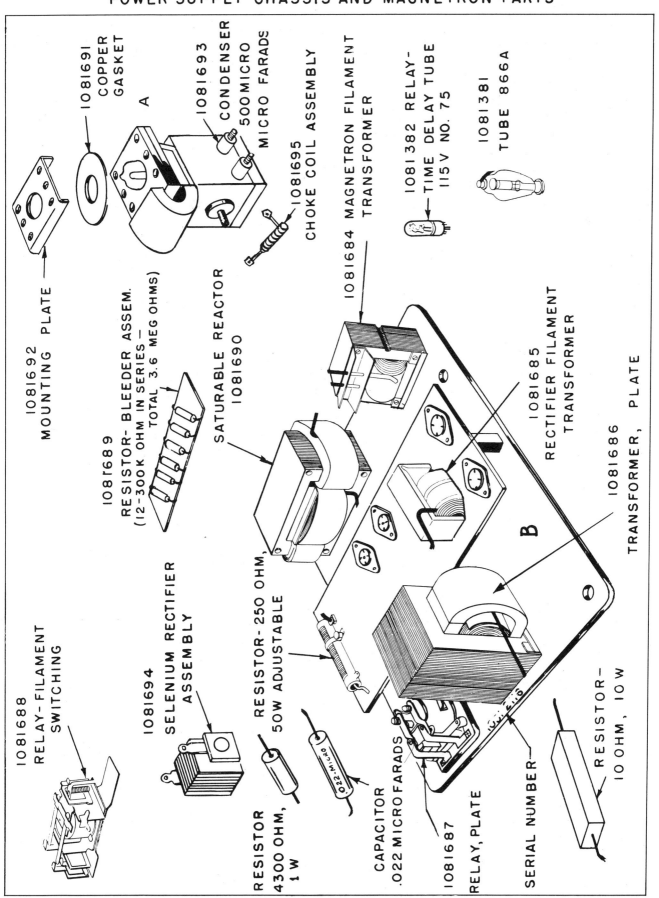

Figure 25 — Power Supply-Magnetron Parts

SECTION 4 EARLY MICROWAVE OVENS

SERVICE PROCEDURE COMPONENT DATA

The following diagnosis and procedures are typical of most early models of microwave ovens.

SERVICE DIAGNOSIS – SECTION

There is 4800 volts of D.C. current rated at 1/3 ampere exposed during the installation, adjustment and service of the microwave oven. Service to the magnetron or power pack should be performed with the POWER OFF. Serious injury or death can result if this warning is not heeded. Factory authorized personnel should be called if service is diagnosed as failure of the magnetron or power pack. DO NOT ATTEMPT TO DEFEAT THE PURPOSE OF THE WARNING TAG ON THE REAR PANEL OF THE OVEN.

Perform the complete operational check before attempting service. This will make the trouble shooting procedure easier and more reliable. The seals must be replaced on the back panel whenever service is performed on the rear of the range.

OPERATIONAL CHECK

Set the "Selector" switch (A) to "Hi" and turn the "Off-On" (B) switch to the "On" position. Make the following inspections, *Figure 24.*

1. Through ventilating screen at bottom of range check to see that the power supply cooling fan is rotating, (C).
2. Check to see that a faint light is visible at top of range. This indicates that the magnetron filaments are heating, *Figure 25(A).*
3. Check to see that the stirrer is rotating, *Figure 24(D).*
4. Feel to see that air is coming from vent at top of range.

If any of the above are not correct, refer to Service Diagnosis.

Within approximately 75 seconds after the range is turned "on", the oven light will come on. This normally indicates that the oven is ready to "cook", but another five minutes MUST be allowed at this point, during an initial installation only, before turning oven timer to "on"(E). If this time is not allowed, *Figure 24,* damage can be done to the range.

Remove recipe file drawer below oven.

Insert meter lead into monitoring jack located to the left of fuses with meter set to read 0-500 milliamps D. C.

Place glass container (measuring cup full of water) in oven, set oven timer to one minute and close oven door. Meter should now read between 310 and 320 milliamps. If not, the following adjustment should be made.

1. Remove fastener to left of fuses, to expose adjustment.
2. Insert screw driver in opening. Turn screw clockwise to increase, counterclockwise to decrease milliamp current, *Figure 24(F).*
3. If screw driver does not provide proper current adjustment (310 to 320 milliamps), the following adjustments must be made:

   a. Reset screw driver adjustment to center.
   b. Disconnect range from power source.
   c. Remove range from cabinet opening and place on suitable support.
   d. Remove seals and back panel. Remove right back clip of Kay Gray material and slide material upward on magnetron filament wires.
   e. Remove the four 866-A rectifier tubes and 75-second time delay tube.
   f. Slide-wire resistor is accessible at this time.
   g. Loosen knurled adjustment knob, and slide band to left to increase, right to decrease. Move approximately 1/8" for each 10 milliamps required.
   h. Reverse above steps and monitor magnetron current again.
   i. Repeat adjustment if necessary.

4. After proper adjustment has been made, turn "hi-Lo" switch to "Lo". Meter should read between 240 and 280 milliamps to indicate switch is working (no adjustment is made on "Lo").

OVEN CALIBRATION CHECK

To adjust oven cooking power to the established standard, proceed as follows: (Do *not* use *metal* containers in oven.)

SECTION 4 EARLY MICROWAVE OVENS

SERVICE PROCEDURE COMPONENT DATA

1. Use a one liter polyethylene graduate and a Weston dial type centigrade thermometer, with scale of −10°C to 100°C.
   a. Turn selector switch to "Hi".
   b. Fill the measure carefully to the one liter mark with a mixture of hot and cold water to obtain a water temperature of approximately 20 to 25 degrees centigrade (°C.).
   c. Stir with the thermometer until well mixed. Read the temperature in degrees centigrade. Record the reading on a piece of paper.
   d. Place the measure of water in the range oven and *heat for exactly two (2) minutes* using a stop watch or the sweep second hand of a good watch.
   NOTE: *Set oven timer at greater than two (2) minutes.*
   e. At end of two (2) minutes cook period, turn oven timer off and quickly remove the measure of water from the oven and again stir thoroughly and measure the temperature. Record this reading also.
   f. Subtract the first reading from the second.
   g. The temperature rise should be 26.5° C plus 1° or minus 2° C.

2. If temperature rise is below 24.5 degrees or above 27.5 degrees, an adjustment should be made by adjusting the magnetron current. Turn screw driver adjustment clockwise to raise and counterclockwise to lower.

3. Repeat power check until proper temperature rise is obtained. NOTE: AT *NO TIME* SHOULD MAGNETRON CURRENT BE SET ABOVE *320* MILLIAMPS.

In case the one-liter polyethylene graduate and centigrade thermometer are not available, use a 1-quart glass milk bottle and a high reading Fahrenheit thermometer (0° to 220°F scale), with 2° increments, similar to Taylor #21418. (An oven tester thermocouple may be used in an emergency to test the water temperature).

An electronic range which has been correctly set for a 310 to 320 milliampere reading will raise the temperature of a 1-quart milk bottle full of 70° to 74°F water, 50° to 55°F in *exactly* two minutes. Stir water thoroughly from top to bottom.

Failure to elevate the water temperature 50° to 55°F may be caused by microwave leakage. Check wave guide screws in ceiling of the oven and magnetron mounting bolts, also, all screws in the oven for looseness.

Check the oven door for seal with a strip of paper two inches wide. The door should clasp the paper all the way around and also near the hinge. Finally, recheck the milliampere setting.

h. OPERATION OF BROILER UNIT

With Off-On" switch "on", set broiler timer to 2 minute setting. Broiler indicator should light and broiler should heat. Allow timer to return to zero, light will go out, broiler element will turn "off" and bell will ring.

SECTION 4 EARLY MICROWAVE OVENS

SERVICE PROCEDURE COMPONENT DATA

SERVICE DIAGNOSIS

DANGER — HIGH VOLTAGE

PROBLEM	POSSIBLE CAUSE	CORRECTION
A. Oven "dead" when "Off-On" switch is turned "on".	1. No power from service entrance. 2. "Off-On" switch inoperative.	1. Check service entrance box and fuses. 2. Replace.
B. Stirrer fails to operate *(Figure 24(D)*.	1. Drive belt off. 2. Loose stirrer pulley. 3. Stirrer off shaft. 4. Motor leads disconnected or broken. 5. Inoperative motor.	1. Replace. 2. Tighten. 3. Replace. 4. Repair. 5. Replace.
C. Power supply cooling fan fails to operate *Figure 24(C)*.	1. Broken lead wire or loose connection. 2. Blade loose on shaft. 3. Inoperative motor. 4. Blade obstructed by cable to power supply.	1. Repair. 2. Tighten set screw. 3. Replace. 4. Reposition.

D. No visible shadow in oven during initial 75 second warm up period (CAUTION: *do not attempt to operate range in this condition.*)	1. Blown fuse, *(Figure 24(F)*.
	2. Loose connections between magnetron and filament transformer.
	3. Inoperative magnetron.
		1. Replace.
		2. Check and tighten.
		3. To check:
		  a. Disconnect range from source of power at service entrance.
		  b. Disconnect magnetron leads from filament transformer.
		  c. Make continuity check of magnetron filaments.
		  d. If it is open, replace magnetron assembly.

NOTE: *If above three possible causes check out O.K., the power supply chassis assembly is at fault and should be repaired.*

SECTION 4 EARLY MICROWAVE OVENS

SERVICE PROCEDURE COMPONENT DATA

PROBLEM	POSSIBLE CAUSE	CORRECTION
E. Oven lamp fails to come "on", *Figure 24(H)*.	1. Blown fuse. 2. Oven lamp burned out. 3. Inoperative time delay tube, *Figure 24(I)*.	1. Replace. 2. Replace. 3. Replace.
F. Oven fails to cook, make complete operational check first.	1. Burned out 30-A tubular fuse in range indicated by no current on meter reading, *Figure 24(J)*. 2. Inoperative 866-A rectifier tube or tubes indicated by very low current meter reading, *Figure 24(K)*. 3. Inoperative oven timer or bad timer wiring. No current on meter reading. 4. Inoperative oven door interlock switch or bad wiring. No current on meter reading, *Figure 24(L)*. 5. Loose magnetron connections. (If arcing occurs, there will be spurts of high current.) 6. Inoperative magnetron. Low current, arcing or excessive current, *Figure 24(M)*. 7. Inoperative power supply chassis. Low current or arcing, *Figure 25(B)*.	1. Replace. 2. CAUTION: *Disconnect range from power and with insulated screwdriver, ground tube caps to chassis. Then replace inoperative tube.* 3. Replace or repair. 4. Replace or repair. 5. Clean and tighten all magnetron connections. 6. Replace. 7. Repair inoperative component.

SECTION 4 EARLY MICROWAVE OVENS

SERVICE PROCEDURE COMPONENT DATA

PROBLEM	POSSIBLE CAUSE	CORRECTION
G. Arcing. (Magnetron or power pack).	1. Non-starting magnetron due to: a. Loose filament connections. b. Improperly timed time delay tube.	1. a. Fasten securely. b. First check 75 second timing of time delay tube. (Wait approximately 5 minutes for magnetron and power supply to cool sufficiently to make this check.) If check discloses 60 to 90 seconds, delay tube is O.K. c. Disconnect power at source, ground rectifier tubes to chassis and manually check relay contacts. Replace relay if necessary. d. Replace.

NOTE: *If arcing resulted in power supply due to the magnetron, a visual inspection of power supply components should be made to assure no permanent damage to power supply has occurred.*

If arcing resulted in magnetron filter assembly due to power pack, a visual inspection of filter assembly should be made.

H. Arcing In Oven.
May be caused by looseness of any metallic oven component or improper door seal.

All door assembly screws exposed to inside of oven should be securely fastened, screws fastening broiler element to sides and back of oven should be tight, and welds connecting broiler element to mounting rod should be secure.

PROBLEM	POSSIBLE CAUSE	CORRECTION
I. Range continues to cook with door open.	1. Inoperative interlock switch, *Figure 24(L)*.	1. Replace switch.
J. Range continues to cook after timer has returned to zero.	2. Inoperative timer, *Figure 24(E)*.	2. Repair or replace.

NOTE: *If both I and J conditions exist, relay is inoperative and must be replaced, Figure 24(N).*

SECTION 4 EARLY MICROWAVE OVENS

SERVICE PROCEDURE COMPONENT DATA

PROBLEM	POSSIBLE CAUSE	CORRECTION
K. Milliamp check of "Hi-Lo" Selector, *Figure 24A*.		
If milliammeter reads between 310 and 320 in both Hi and Lo positions on selector switch, replace switch.		
If milliammeter reads approximately 280 in both Hi and Lo positions on selector switch, replace switch.		
If milliammeter reads between 310 and 320 in Hi position on selector switch and reads below 100 milliamps on "Lo" position, replace resistor (100 ohms and 25 watts) which is mounted beside Hi-Lo switch.		
L. Broiler element indicator light fails to come on.	1. Range "Off-On" switch in "off" position or defective, *Figure 24(B)*. 2. Burned out lamp. 3. Loose wire. 4. Inoperative broiler timer.	1. Turn on or replace. 2. Replace. 3. Repair. 4. Replace.
M. Broiler element fails to heat with indicator light on.	1. Element over-heated and safety limit switch open. 2. Inoperative safety switch. 3. Loose or broken connections. 4. Inoperative broiler unit. 5. Inoperative broiler timer.	1. Allow 10 to 15 minutes for unit to cool and reset timer. 2. Check for continuity when cool, if defective, replace. 3. Repair. 4. Replace. 5. Repair or replace.

SECTION 4 EARLY MICROWAVE OVENS

SERVICE PROCEDURE COMPONENT DATA

N. **Fuses continue to blow.**

1. 30 amp, *Figure 24(J)*.
   Insert milliammeter lead into monitoring jack and replace blown fuse before turning on oven timer. Observe meter reading at instant timer is turned on.

   CAUTION: *If current is excessive, meter may read above 500 milliamps and timer should be turned off immediately to avoid damage to milliameter or range.*

   First, exchange magnetron. If condition still exists, replace magnetron with original and locate trouble in power pack and replace faulty component.

2. 0.6 or 3.2 amp, *Figure 24(G)*.
   If either the 0.6 amp or 3.2 amp fuses continue to blow, check to see that the range is properly connected to single phase 3-wire 220 volts. If wiring is correct, check for a short circuit in the power source.

   Milliamp load above 320 will blow 3.2 amp. fuse.

O. **Power check shows very low temperature rise of water.**

   Check for possible arcing in oven or at magnetron or wave guide. If arcing is between magnetron and matching section, or on magnetron, change magnetron plate and gasket.

P. **Timer Problems,** *Figure 24(E)*.

1. Sticking:
   If a timer has been sticking during operation, it is possible that either the timer shaft or the dial are binding. Check to see if the shaft is centered in the hole in the control panel. If it is not, loosen mounting screws and adjust. Then, tighten screws firmly. Tightening these screws will normally correct the trouble of the dial binding on the timer knob back plate, but as a last resort, a shim such as friction or plastic tape, can be inserted in the hollow shank of the knob to relocate the dial away from the back plate.

2. Switch does not turn on, *Figure 24(B)*.
   Check first to determine if switch or interlock is at fault. The actuator arm which operates the plunger on the switch could be bent and not move plunger far enough in to close the switch. This arm can be bent slightly toward the plunger with a long nose pliers enough to cause switch to operate. Care must be taken so that the arm is not bent so far that switch will not turn on.

3. Switch does not turn off.
   The actuator arm which operates the plunger on the switch could be bent enough that the switch will not turn off. This arm can be bent with long nose pliers enough to allow plunger of switch to operate, but care must be taken so that arm is not bent so far that switch will not turn on.

   The knob on the face of the timer, which is attached to the actuator arm may be pressed against the control mounting plate and not allow the arm to return to the "off" position when the timer has returned to the zero position. Reposition the timer so that this knob is free from any drag at this point.

SECTION 4 EARLY MICROWAVE OVENS

SERVICE PROCEDURE COMPONENT DATA

REPLACEMENT OF PARTS — SECTION

a. OVEN TIMER, *Figure 24(E)*.

To remove and install: (Disconnect range from power source)

1. Pull timer knob off.
2. Remove two bezel screws and lift out control panel assembly as far as wire slack will permit.
3. Remove two machine screws which fasten timer assembly to control panel mounting plate.
4. Transfer one wire at a time from inoperative timer to corresponding terminal on new timer.
5. Mount new timer to mounting plate.
6. Replace control panel assembly, back plate and knob.

b. OFF-ON SWITCH, *Figure 24(B)*.

To remove and install: (Disconnect range from power source)

1. Pull switch knob off.
2. Remove two bezel screws and lift out control panel assembly as far as wire slack will permit.
3. Remove hex nut from back plate.
4. Transfer one wire at a time from inoperative switch to corresponding terminal on new switch.
5. Mount switch to control panel and secure with nut.
6. Replace control panel assembly and knob.

c. HI-LO SWITCH, *Figure 24(A)*.

To remove and install: (Disconnect range from power source)

1. Follow same procedure as for "Off-On" Switch, Item (B.)

d. BROILER TIMER

To remove and install: (Disconnect range from power source)

1. Follow same procedure as for "Oven Timer", Item a.

e. BROILER INDICATOR LIGHT

To remove and install: (Disconnect range from power source)

1. Remove two bezel screws and lift out control panel assembly as far as wire slack will permit.
2. Push bulb assembly forward to disengage from control panel bezel, disconnect wires and remove.
3. Thread lead wires of new bulb assembly through control panel, press bulb assembly into opening and connect leads to corresponding locations.

f. OVEN INDICATOR LIGHT

To remove and install: (Disconnect range from power source)

1. Follow same procedure as for Broiler Indicator Light, Item e.

g. CONTROL PANEL BEZEL

To remove and install: (Disconnect range from power source)

1. Remove all controls, both indicator lights and two screws which fasten bezel and mounting plate to front frame.

h. FRONT OUTER FRAME

To remove: (Disconnect range from power source)

1. Pull oven assembly forward approximately four (4) inches from cabinet.
2. Remove recipe file drawer.
3. Remove two pins from inside of recipe file drawer opening. Loosen two metal screws from screen, at end nearest front frame. Remove screen.
4. Remove all screws which fasten front frame to oven wrapper.

SECTION 4 EARLY MICROWAVE OVENS

SERVICE PROCEDURE COMPONENT DATA

    a. Five screws inside oven door opening.
    b. Two screws in screen opening.
5. Remove two bezel screws and lift control panel assembly forward.
6. To disengage front frame from magnetron cooling channel, lift front frame forward at bottom and up.

To install:

1. Reverse all steps, (1) through (6).

i. OVEN DOOR ASSEMBLY

To remove:

1. Pull range forward approximately two (2) inches from cabinet opening.
2. With a #3 Phillips screw driver, loosen oven door spring screws (at lower right and left front corner of base) and remove all piano hinge screws, then lift off door withdrawing left and right hinge cams through slots in front frame.

To install:

1. Insert hinge cams through front frame and fasten piano hinge to frame.
   NOTE: *Make certain door liner seals flush at all surfaces of oven front when door is closed.*
2. Tighten spring screws.
3. Replace range in cabinet opening.

j. OVEN DOOR SPRINGS

To remove and install:

1. Disconnect range from power source.
2. Remove oven from cabinet opening and place on suitable support, such as a padded table.
3. Remove panel from side of range containing defective spring.
4. Remove #3 Phillips screws from spring and replace spring.
5. Replace panel, install oven and connect to power source.

k. OVEN DOOR ROLLER BRACKET ASSEMBLY

To remove or install bracket or roller:

1. Disconnect range from power source and remove oven door spring.
2. Remove two (2) bolts and nuts from above and below slots in front frame.
3. Replace parts, install oven and connect to power source.

l. DOOR SAFETY SWITCH (In power relay circuit)

To remove and install:

1. Disconnect range from power source.
2. Remove range from cabinet opening and place on padded table.
3. Remove panel from side of range containing inoperative interlock switch, and open oven door.
4. Remove switch from mounting and disconnect wires.
5. Install new switch.
6. Carefully close oven door to check for contact with switch.
   CAUTION: *Excessive pressure from hinge cam will damage switch.*

m. OVEN LAMP

To install: (Disconnect range from power source)

1. Remove control panel assembly.
2. Loosen one screw and remove cover at front of light box.
3. Unscrew burned out lamp and replace.

Oven Lamp Socket Replacement:

1. Disconnect range from power source.
2. Remove control panel assembly.
3. Remove two (2) hex nuts which fasten oven light box to top of oven.
4. Remove oven light box cover (two (2) screws).

SECTION 4 EARLY MICROWAVE OVENS

SERVICE PROCEDURE COMPONENT DATA

5. Unscrew lamp and ceramic ring from socket.
6. Remove wires and transfer to terminals of replacement.
7. Re-assemble by reversing steps (1) through (6).

n. **HI-LIMIT SWITCH**

To remove and install:

1. Disconnect range from power source.
2. Remove control panel assembly.
3. Remove two (2) bolts and nuts which fasten switch to top of oven.
4. Disconnect wires and install new switch.

o. **STIRRER**

To remove and install:

1. Open oven door.
2. Remove glass.
3. Remove Phillips screw and shakeproof washer from stirrer shaft and lift out.
4. Replace in reverse order.

p. **STIRRER ASSEMBLY**

To remove and install:

1. Disconnect range from power source.
2. Remove range from cabinet opening and place on padded table.
3. Remove top and back panel.
4. Remove stirrer drive belt from stirrer pulley.
5. Remove stirrer.
6. Remove four (4) bolts and nuts from stirrer bearing housing and withdraw stirrer assembly.
7. Remove stirrer pulley and install on replacement stirrer assembly.
8. Reverse above procedure, steps (1) through (6).
9. Do not oil the stirrer bearing.

q. **MAGNETRON**

To remove and install:

1. Disconnect range from power source.
2. Remove range from cabinet opening and place on padded table.
3. Remove top and back panel and ground tube caps to chassis with insulated screw driver.
4. Remove two wing nuts and disconnect magnetron leads from filter assembly.
5. Fold rubber sleeving back on itself and the blower assembly.
6. Remove rubber filter cooling tube by pulling from openings in filter assembly blower housing.
7. Remove drive belt from blower motor pulley and remove motor lead wire from twist on connector.
8. Remove three (3) screws from underside of bracket which fasten into bottom of blower housing and lift off assembly.
9. With a non-magnetic (*this is important*) 7/16" open end wrench, remove the four (4) hex nuts which fasten the magnetron to the bottom side of wave guide. NOTE: *Magnetron assembly will remain suspended in a secure position due to locking pins located in tubular brass bushings between four (4) magnetron mounting bolts.*
10. Grasp magnetron at each side and with thumbs, depress locking pins toward each other to unlock and carefully lower magnetron from opening in bottom of wave guide to avoid breaking magnetron tube.
    CAUTION: *Place magnetron assembly in special magnetron shipping container to avoid damage.*
11. When replacing magnetron, locking pins will automatically lock assembly in place to facilitate replacing four (4) nuts on mounting plate.
12. Reverse above procedure, steps (1) through (10), with special attention to rubber sleeve and rubber tube installation.

r. **WAVE GUIDE**

To remove and install:

1. Disconnect range from power source.
2. Remove magnetron (see Item "q"), and stirrer assembly (see Item "p").
3. Remove 14 machine screws that fasten

SECTION 4 EARLY MICROWAVE OVENS

SERVICE PROCEDURE COMPONENT DATA

wave guide to oven top from inside of oven.
4. Install replacement, being careful to start all machine screws and tighten evenly and securely. *This is important* to obtain a good seal.
5. Re-assemble magnetron and stirrer as per Items "p" and "q".

s. **MAGNETRON BLOWER MOTOR**

To remove and install:

1. Disconnect range from power source.
2. Remove range from cabinet opening and place on padded table.
3. Remove top and back panel.
4. Remove rubber sleeve from magnetron and blower housing.
5. Remove motor lead wire from dis-connect and ground.
6. Remove stirrer drive belt.
7. Remove three (3) screws from blower cover and lift off motor with cover.
8. With 1/8" Allen wrench, remove set screw and fan assembly.
9. Remove four (4) nuts that fasten motor to cover.
10. Transfer friction drive pulley and post to replacement motor assembly.
11. Replace fan on shaft. Check clearance.
12. Replace motor assembly to blower cover.
13. Place rubber sleeve on blower housing and fold back on itself to facilitate attaching to magnetron cooling fins.
14. Reverse steps (1) through (6).

t. **POWER SUPPLY CHASSIS (Power Pack)**

To remove and install:

1. Disconnect range from power source.
2. Remove range from cabinet opening and place on padded table.
3. Remove back panel only and ground tube caps to chassis with insulated screw driver.
4. Disconnect two (2) lead wires which fasten magnetron to power supply.
5. Remove two wing nuts (one each rear corner), which fasten power supply back to base of range.
6. *Carefully* lift out power pack as far as wires will permit.
7. Disconnect lead wires (pull off) and remove power pack.
8. Position replacement power supply at rear of range.
9. Replace leads to corresponding terminals and ground lead to base of power pack.
10. Transfer 866-A and time delay tubes to replacement power pack.
11. As power pack is placed in base of range, visually inspect and clear all wires.
12. Reverse above procedure, steps (1) through (5), and make certain filament leads to magnetron are securely fastened.

Before attempting to replace any of the components of the power supply chassis (power pack), remove the chassis from the range and place it on a work bench. Remove all the tubes. Tag all wires removed so that they will not become mixed.

Saturable Reactor

To remove:

1. Unsolder leads from reactor to selenium rectifier assembly at the rectifier assembly.
2. Unsolder large black leads on top of reactor.
3. Stand chassis on end so that it rests on the plate transformer.
4. Unfold and straighten clamping tabs which mount reactor to chassis.
5. Remove reactor.

To install:

1. Reverse the above procedure.

Plate Transformer

To remove:

1. Unsolder leads from transformer at the base of the tubes.
2. Unsolder leads at primary.
3. Stand chassis on end and straighten clamping tabs which mount transformer to chassis.
4. Remove transformer.

SECTION 4 EARLY MICROWAVE OVENS

SERVICE PROCEDURE COMPONENT DATA

To install:

1. Reverse the above procedure.

Magnetron Filament Transformer

To remove:

1. Slide insulating sleeves back from solder connections at primary and unsolder.
2. Stand chassis on end and straighten clamping tabs which mount transformer to chassis.
3. Remove transformer.
4. Transfer bleeder assembly and wire with tube clamps to new transformer.

To install:

1. Reverse steps (1) through (3) above.

866A Filament Transformer

To remove:

1. Remove screw securing ground terminal.
2. Remove (2) screws securing power relay.
3. Stand chassis on end and straighten clamping tabs that mount fiber board panel to chassis.
4. Remove screw that holds fiber board panel to ceramic spacer.
5. Lift fiber board panel and power relay to expose connections from transformer to tubes.
6. Unsolder connections at tube sockets.
7. Drill out rivets mounting transformer to fiber board.
8. Remove transformer.

To install:

1. Reverse above steps using machine screws and nuts instead of rivets to secure transformer to fiber board.

Selenium Rectifier Assembly

To remove:

1. Unsolder leads at top of selenium rectifier assembly.
2. Remove screw securing ground terminal.
3. Remove (2) screws securing power relay.
4. Drill out rivet securing right selenium rectifier mounting bracket.
5. Stand chassis on end and strighten clamping tabs that mount fiber board panel to chassis.
6. Remove screw that holds fiber board panel to ceramic spacer.
7. Lift fiber board panel and power relay to expose selenium rectifier assembly.
8. Pry off tinnermans holding rectifier assembly to mounting brackets.

To install

1. Reverse above steps using a machine screw and nut instead of a rivet to mount the right hand mounting bracket.

Filament Relay

To remove and install:

1. Unsolder all wire connections and transfer to new relay.
2. Drill out rivets which fasten relay to fiber board panel.
3. Fasten new relay to panel with nuts and bolts instead of rivets.
4. Transfer condenser and resistor to new relay.

Power Relay

To remove and install:

1. Transfer all wire connections to new relay.
2. Remove two screws in base of relay.
3. Install new relay.

Bleeder Resistor Assembly

To remove:

1. Remove screw fastening wire from transformer lead to bleeder assembly.
2. Remove nut fastening assembly to transformer.
3. Remove above assembly.
   (a) If only one resistor is bad on this assembly, replace only this resistor.

SECTION 4 EARLY MICROWAVE OVENS

SERVICE PROCEDURE COMPONENT DATA

To install:

1. Reverse the above procedure.

Slide Wire Resistor

To remove:

1. Unsolder on one end of resistor.
2. Remove slider from resistor by removing nut which fastens it to resistor.
3. Pry out mounting spring from one end of resistor.

To install:

1. Reverse the above procedure.

u. 866-A RECTIFIER TUBES AND TIME DELAY TUBE

To replace:

1. Disconnect range from power source.
2. Remove range from cabinet opening and place on padded table.
3. Remove back panel and ground tube caps to chassis. (With insulated screw driver).
4. Remove defective tube and install replacement. NOTE: *Whenever 866-A tube is replaced, 5- minute warm-up period is required before turning oven timer on. (To properly condition tube).*
5. Replace parts in reverse order, steps (1) through (3).

v. BROILER ELEMENT

To remove and install:

1. Disconnect range from power source.
2. Remove 12 screws that mount broiler element in oven.
3. Pull element forward *carefully* and remove slip on connectors. CAUTION: *If these leads are released, they may slip back through opening, making it necessary to remove oven from opening.*
4. Install replacement by reversing steps (1) through (3) above. NOTE: *Make certain mounting screws are tight to avoid arcing.*

w. COOLING FAN ASSEMBLY (Power Pack)

To replace:

1. Disconnect range from power source.
2. Remove recipe file drawer and remove screen.
3. Cut off leads close to motor mounting frame.
4. Remove motor assembly and bracket and lift out.
5. Transfer replacement motor assembly to bracket.
6. Install bracket and motor assembly.
7. Use twist-on connectors or suitable fastening on motor leads.
8. Replace screen and file drawer.
9. Connect to power source.

x. RANGE FUSE LOCATION

0.6- and 3.2-amp fusestats accessible through recipe file drawer. 30-amp tubular fuse accessible through rear of range by removing back panel.

SECTION 4 EARLY MICROWAVE OVENS

SERVICE PROCEDURE COMPONENT DATA

SCHEMATIC WIRING DIAGRAM

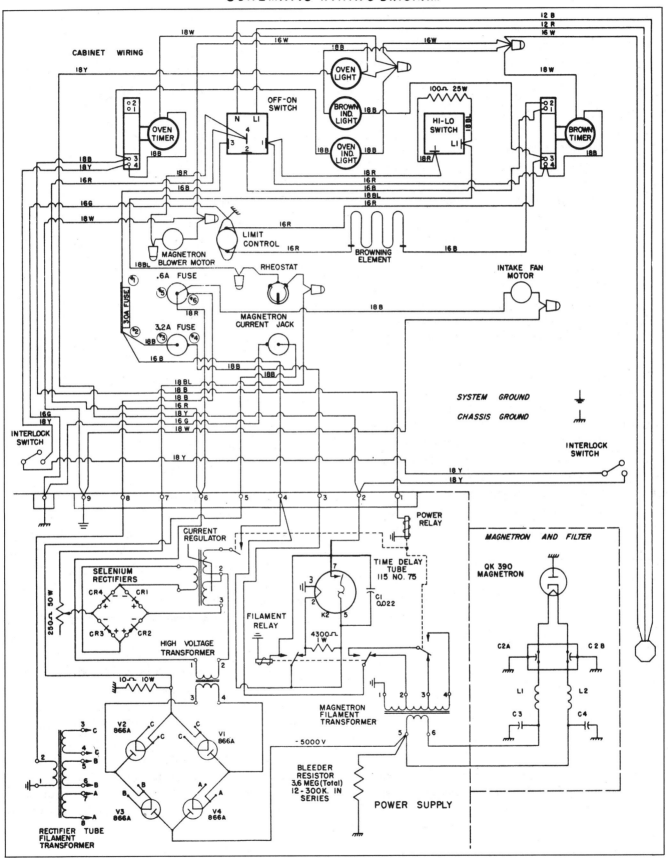

SECTION 5 AMANA RADARANGE

SERVICE PROCEDURE COMPONENT DATA

INTRODUCTION

Microwaves are electromagnetic waves of energy, similar to radio, light, and heat waves. This is a general definition and covers a number of familiar types of radiation that are not normally considered microwaves, such as radio, TV and infra-red. To make the definition more specific, the limitation that the wave length of the radiant form of energy must lie somewhere between 1 meter and 1 millimeter. Radio and TV are thus excluded because their wave lengths are much longer, and infra-red is excluded because its wave length is much shorter.

There is a direct relationship between wave length and frequency. Wave length is equal to the speed of light divided by the frequency. For example, let's take a radio station which operates at 600 kilocycles. The wave length would, therefore, be the speed of light, 3 million meters per second, divided by the frequency 600,000 cycles per second which is equal to approximately 500 meters.

The Federal Communication Commission limits or controls the design of microwave ovens. They have allocated a number of frequency bands that may be used for the operation of a microwave oven and some other types of related equipment. The first of these frequencies is 915 megacycles, the second is 2450 megacycles, and a still higher one is 5800 megacycles. There are a few allocations at higher frequencies, but these are not generally used in microwave ovens and are, therefore, not worth mentioning. Almost all of the microwave ovens that are produced today operate at 2450 megacycles.

Microwaves have many of the same characteristics that light waves have.
1. They travel in a straight line.
2. They can be generated.
3. They can be reflected, transmitted and absorbed.

The basic differences are what materials reflect, transmit and absorb, and how the microwaves are generated, *Figure 26*.

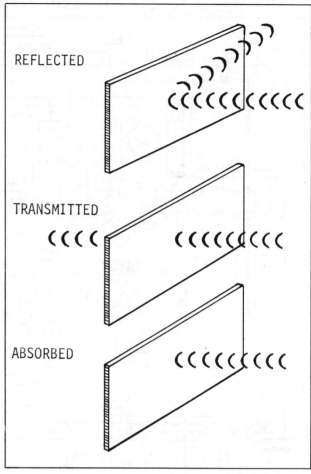

Figure 26

The simplest and the most commonly used generator for producing light is a light bulb. In a microwave oven, the generator for producing the energy is a magnetron. The magnetron is a vacuum tube which operates as an oscillator to generate the microwave energy. Every radio and TV set has an oscillator circuit. This circuit consists of a number of vacuum tubes, resistors, capacitors, and conductors. In the microwave oven, all of these oscillator components are built into the tube.

Although microwaves can be reflected in the same manner that light is, the materials that reflect do vary. For example, aluminum and stainless steel reflect microwaves while cold-rolled steel, to some extent, absorbs the microwave power. Another example is the perforated holes in the door of the microwave oven. These holes do reflect the microwave power, but they transmit light. Opaque paper products transmit microwaves while light waves are either absorbed or reflected. Glass and china act much the same as paper but some do absorb power. Water absorbs microwaves while light passes through.

The selective characteristics of microwaves make it possible to construct an oven where the wall, ceiling, floor, cooking container, and door remain cool. These items do get warm, but this is caused by the transfer of heat from the food.

Foods are heated in the oven by the absorption of microwave power. Food is constructed of many millions of molecules per cubic inch. These molecules react to the microwave field much the same manner as a compass needle reacts to a magnet. If you place a magnet to one side of the compass, the needle will then point to the magnet. If you then move the magnet to the other side of the compass, the needle will turn and again point to the magnet. When this process is repeated quickly and many times, eventually the friction in the bearing that supports the needle will be heated.

The molecules in food react in a very similar manner to the changing microwave field, that is, the molecules tend to align themselves with the field. The molecules that make up the food which is cooked in oven are rotated from their starting position to 180° from their starting position and back to their starting position 2,450 million times a second. This constant and rapid rotation causes the food to heat. As the wave penetrates the food, power is lost to each successive layer of molecules. The center molecules are therefore not rotated a full 180° unless heat is generated towards the center of the food as opposed to the outside of the food, *Figure 27*.

Contrary to popular belief, food prepared with a microwave oven is not cooked from the inside out, but is cooked all the way through at the same time with more cooking being performed on the exterior of the food. It is, therefore, possible to prepare a rare, medium, or well done roast in this type of oven.

The fact that food is heated throughout makes it possible to cook food fast. Time required to cook an item is solely dependent upon how much heat is required, and in turn, the amount of heat required of the food and the weight of the food. In the conventional process, only the surface of the food is heated directly by the oven or grille. The heat required to cook the inside portion has to be conducted from the surface. Three factors govern the time required to cook an item in the more conventional way. A minor one of these is how much heat is required. The major ones are, how well does the food conduct heat, and how much can the surface be overheated without causing serious defects. For example, let's take water. Water is a good conductor of heat and the surface can be overheated without deterioration, therefore, water can boil fairly quickly on a range. On the other hand, let's take milk. Milk is also a good conductor, but the surface cannot take overheating. If you try to boil milk quickly, the milk will burn, therefore, milk has to be cooked slowly. Cake is an example of a food that conducts heat poorly. Although a small amount of heat is required to bake a cake, it must be cooked rather slowly because the conduction to the center is poor.

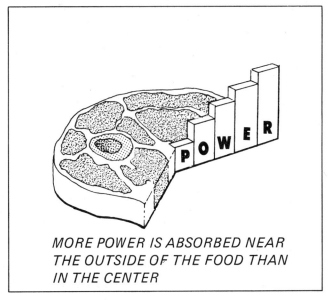

MORE POWER IS ABSORBED NEAR THE OUTSIDE OF THE FOOD THAN IN THE CENTER

Figure 27

MAGNETRON, *Figure 28*
1. Output antenna
2. Magnetron pole pieces
3. Strap rings
4. Vanes
5. Anode block
6. Heater and cathode
7. Magnetron housing
8. Magnet coil

SECTION 5 AMANA RADARANGE

SERVICE PROCEDURE
COMPONENT DATA

9. Magnet pole
10. Cooling fins
11. Heater leads and cathode lead
12. Glass
13. Wave Guide

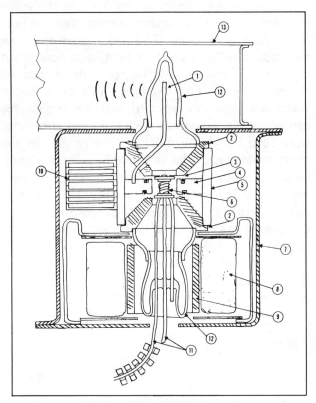

Figure 28 — Magnetron Parts

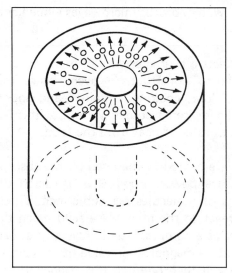

Figure 29 — Heater and Cathode

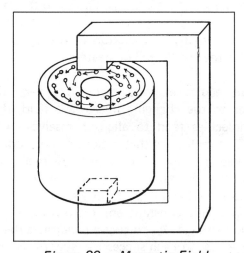

Figure 30 — Magnetic Field

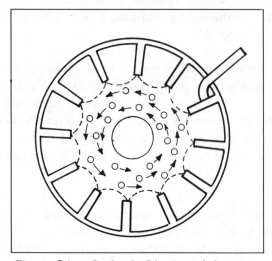

Figure 31 — Cathode Block and Antenna

The inner cylinder of *Figure 29* represents the heater and cathode. The outer cylinder represents the anode block, strap ring and vanes. Heating the center cylinder "boils" electrons free. By imposing a high voltage between the two cylinders, electrons start to travel from the inner to the outer cylinder.

A magnetic field is imposed parallel to the cylinder, *Figure 30.* The magnetic field causes the electrons to spin about the inner cylinder. The "C" shaped magnet in *Figure 30* represents the following components: magnetron pole pieces, magnetron housing, magnet coil, and magnet pole. Items marked 2, 7, 8 and 9 on magnetron illustration *Figure 28.*

Figure 31 is a top view of the cathode, anode block, and antenna. The spinning electrons between the anode and cathode generate the microwave power which is in turn "picked off" by vanes and in turn transmitted from the magnetron by the antenna.

62

SECTION 5 AMANA RADARANGE

SERVICE PROCEDURE COMPONENT DATA

MICROWAVE OVENS

Microwave Generator and Related Components.
The heart of the microwave oven is a magnetron or "maggie" for short. Before World War II, microwave energy was nothing more than a laboratory curiosity. Not too many people understood it and fewer knew how to produce it. However, from this laboratory beginning it was found possible to send out a pulse of microwave energy and if it struck an object, it would bounce off and come back to the sending set. This was the start of radar. The tremendous possibilities for radar in wartime use was foreseen by the people in the armed services. An all-out effort was started to find a better means to produce microwave energy. The quest was answered by the "maggie".

Raytheon became one of the leaders in this design and production of magnetrons. During World War II, Raytheon produced more magnetrons than all the other United States companies combined. In fact, more than twice as many as the nearest competitor.

After World War II, Raytheon embarked on a program to find a new use for microwave energy. One use of the microwave was in the kitchen. The first heating units were produced in 1945. The units were not available for sale but merely to use for testing purposes. In 1954, Raytheon produced the first practical microwave oven.

The magnetron, with the help of the transformer, rectifier, and magnetic field circuit, converts 60 cycle power into microwave power.

Transformer

The transformer is by far the heaviest component in the microwave oven. It is constructed of iron and copper wire. The transformer has a two-fold function — one is to step the household voltage up and one is to step the household voltage down. The input is standard household voltage or 115 volts. The one secondary voltage is 3.1 volts AC and the other is about 4000 volts AC. The 3.1 volt AC is used to heat the filament of the magnetron. The high voltage is connected to a rectifier.

Rectifier

The rectifier is a solid state device which can convert the 4,000 volts alternating current into 4,000 volts DC. The negative pole of the rectifier is connected to a magnetron and the positive end is connected to a magnetic field circuit.

Magnetic Field Circuit

The major component of the magnetic field circuit is a magnet coil. The magnet coil in conjunction with the tube support, which is a sheet metal part, provides the magnetron with an extremely strong magnetic field. In addition to the coil and tube support, there are three minor components which comprise the circuit. One is a capacitor which smooths out the DC power supply to the coil and, therefore, smooths out the magnetic field. The other two are resistors which are used to adjust the complete circuit.

Magnetron

The magnetron, with the help of the transformer, rectifier, and magnetic field circuit, converts the household 60 cycle electricity into microwave energy. The magnetron is made up of a few major components — the input leads to which the heater and high voltage connections are made, the main body of the tube where the microwave energy is created, cooling fins to cool the tube, and an antenna to radiate the microwave power from the magnetron, *Figure 32*.

Blower

Some heat is generated within the tube and this heat has to be removed by forced air. Therefore, the unit is provided with a blower. The blower intake is located in the bottom of the unit and the air is exhausted out the back of the unit. The air intake is provided with a filter. The filter may be removed from the bottom of the unit and cleaned. Since the quantity of air is small, the filter need not be cleaned more than once a year.

Control Circuit.

Since the microwave cooking process is dependent on time, the major components of the control circuit are two timers. In addition to the two timers there are a number of switches.

SECTION 5 AMANA RADARANGE

SERVICE PROCEDURE
COMPONENT DATA

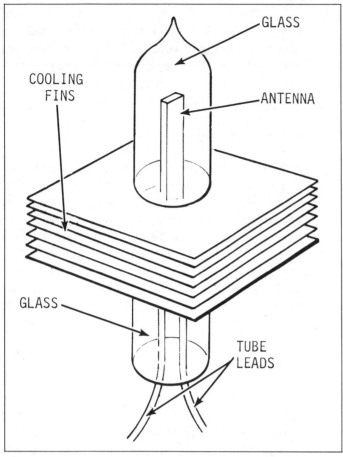

Figure 32 — Magnetron Assembly

Timers

The most often used timer is the bottom timer which has a time scale from 0 to 5 minutes. Approximately 70% of the items cooked will be cooked in less than 5 minutes. Some of the items that will be cooked in this time are hamburgers, hot dogs, bacon and reheating of food. A second timer is provided for cooking longer term items such as chicken and roasts. The time scale on this timer is from 0 to 25 minutes. The reason for providing two timers is that certain cooking items such as hot dogs, which require 45 seconds, must be set very accurately. It would be almost impossible to set 45 seconds on a timer which had a full scale of 25 minutes. The timing circuit is so designed that only one timer will operate at a time.

Door Interlock Switches.

On either side of the door, a door interlock switch is located. These switches stop the unit as soon as the door seal starts to break. No microwave power is emitted when the door is open. The magnetron stops emitting microwave power with the opening of the power circuit.

Start Switch.

The unit will not operate unless one of the timers is set and the start switch is depressed. The start switch is returned to the "off" position as soon as the door is open. The unit may be either shut-off by turning one of the timers to zero or opening the door.

MICROWAVE POWER TRANSMISSION COMPONENTS.

From the magnetron the microwave energy must be directed to the food. One of the basic design considerations which has to be taken into account is that when the energy reaches the food it will be heated evenly. Since microwaves react much the same way as light and travel in a straight line, the understanding of these components is quite simple.

Wave Guide.

The magnetron is connected to a wave guide. A wave guide is a rectangular piece of tubing approximately 2" by 4". The wave guide channels the power from the magnetron into the oven. One end of the wave guide is open and this end serves as an antenna to radiate the power into the oven cavity, *Figure 33*.

Stirrer.

Immediately in front of the antenna is a fan. The microwave energy emitted by the antenna strikes the fan which is rotated at a slow RPM. The fan in turn reflects the power, bouncing it off the walls, ceiling, back, and bottom of the oven. The power enters the food from all sides which gives an even heating pattern in the food, *Figure 33*.

Glass Plate.

Food to be cooked in the microwave oven is placed on a glass plate which is suspended approximately 1" above the bottom of the oven. The glass plate is of special quality Pyrex glass which is transparent to microwaves. The microwaves go through the glass, strike the bottom of the oven, and are reflected back up into the food from the bottom side. This allows the microwave energy to enter the food from all sides, *Figure 33*.

SECTION 5 AMANA RADARANGE

SERVICE PROCEDURE COMPONENT DATA

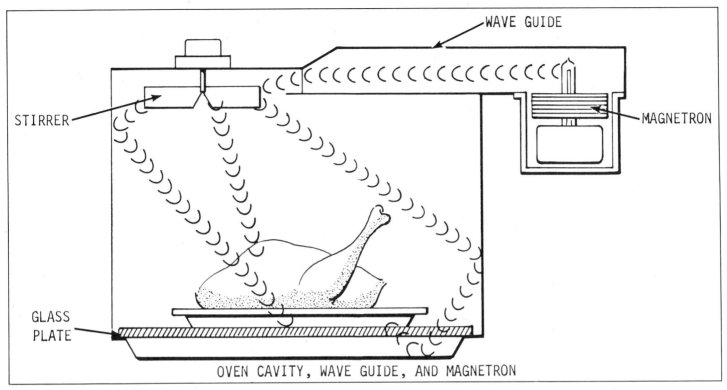

Figure 33 — Microwave Pattern

CONTAINMENT COMPONENTS.

Once generated, the microwaves must be contained so that only the food is cooked. The components are constructed of various materials which reflect, transmit, or absorb power depending on their intended function.

Oven Walls.
The oven enclosure is constructed of stainless steel. Stainless steel reflects microwave energy and does not absorb any power. Aluminum will do the same job, but it does not have the fine appearance as stainless steel, nor is it as easy to clean.

Primary Door Seal (Choke).
The primary microwave seal is the choke. The choke is built on the periphery of the door and fits inside of the oven cavity. Microwaves, like most everything, follow the patch of least resistance, therefore, the microwaves pass by the front edge of the door, strike the back edge of the door, are reflected into the choke cavity, and in turn are reflected back into the oven cavity. The choke cavity is filled with a material which is transparent to microwaves, *Figure 34.*

Secondary Seals.
Any power that by-passes the choke and is not reflected back in the oven is absorbed by the door gasket. The door gaskets are vinyl, but a special grade of vinyl. Ordinary vinyl is transparent to microwaves so the vinyl in the door gasket is loaded with carbon black. This makes the vinyl highly absorbent and, therefore, it has ability to absorb the microwave power which is by-passed by the choke.

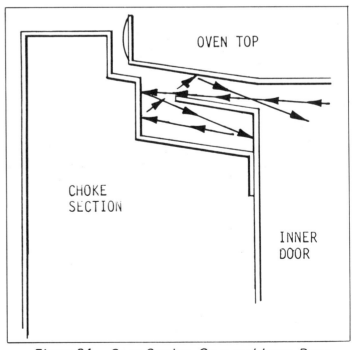

Figure 34 — Cross Section, Oven and Inner Door

SECTION 5 AMANA RADARANGE

SERVICE PROCEDURE COMPONENT DATA

GLOSSARY OF TERMS

AC VOLTAGE
Alternating current.

ANNODE
The positive electrode which receives the electrons in a vacuum tube. In the magnetron, the annode is usually the outer shell and at ground voltage potential.

BRIDGE RECTIFIER
A full way rectifier with four elements connected in series to form a diamond. An AC voltage source is connected between one pair of opposite junctions, and an output is connected between the other pair of junctions. The output voltage is a pulsating DC voltage.

CAPACITANCE
The electric size of a capacitor. The basic unit used is the farad, but the smaller microfarad and picofarad units are commonly used. Also called capacity. The property that exists whenever two conductors are separated by an insulating material, permitting the storage of electricity.

CAPACITOR
A device consisting essentially of two conducting surfaces separated by an insulating material such as air, paper, glass, mica, or plastic film. A capacitor stores electric energy, blocks the flow of direct current, and permits the flow of alternating current to a degree dependent on its capacitance and the frequency.

CAPACITOR, ELECTROLYTIC
Two conducting electrodes with the annode having a metal oxide film formed on it. The film acts as a dielectric or insulation medium. Generally used for filtering, bypassing, coupling, or decoupling. Characteristics: color, high dielectric constant, low voltage, (under 550 volts), low impedance.

CATHODE
A cathode is that portion of a vacuum tube which provides the source of electrons for the tube current. In the magnetron, the cathode is centered within the annode and at a highly negative voltage potential.

CHOKE
An inductance used in a circuit to present a high impedance to frequencies without appreciably limiting the flow of direct current. Also called a coil. A grove or other discontinuity in a wave guide so shaped in dimension as to reflect guided waves within a limited frequency range.

CIRCUIT
A path over which an electrical current can flow.

COIL
A number of turns of wire used to introduce inductance into an electric circuit, to produce magnetic flux or react mechanically to a changing magnetic flux. In high-frequency circuit, a coil may be only a fraction of a turn. The electrical size of a coil is called inductance and is expressed in henrys. The opposition is called impedance and is expressed in ohms. The impedance in a coil increases with frequency. Also called inductance and inductor.

DIELECTRIC
A metal that can serve as an insulator because it has poor conductivity. A dielectric, such as air, mica, or plastic film is used between the metal foil plates of a capacitor to separate the plates electrically and store electric energy. A dielectric undergoes electric polorization when subjected to an electric field.

DIODE
Basic in semi-conductor art. It passes current in one direction and blocks it in the other.

ELECTRON
High-speed, negatively charged particle forming outer shell of an atom, smallest electric charge that can exist.

FERRITE
A powdered, compressed, and sintered magnetic material having high resistivity consisting chiefly of ferrious oxide combined with one or more materials. The high resistance makes any current losses extremely low at high frequencies.

FILAMENT OR HEATER
A wire which, when heated with an electrical current, emits electrons.

SECTION 5 AMANA RADARANGE

SERVICE PROCEDURE COMPONENT DATA

FILTER CAPACITOR
A capacitor used in a power supply filter system to provide a low-reactance path for alternating currents and, thereby, suppresses ripple currents, without effecting direct currents. Electrolytic capacitors are generally used for this purpose.

IMPEDANCE
The total opposition that a circuit offers to the flow of alternating current or any other varing current of a particular frequency. A combination of resistance and reactance, and expressed in ohms.

INDUCTANCE
The property of a circuit that causes a magnetic field to be induced in a circuit by a change in the circuit. The henry is a unit of inductance.

MAGNETRON
A magnetron is a vacuum tube in which the flow of electrons from the heated cathode to the annode is controlled by a magnetic field and an electric field. It is used to produce very short electrical waves which are termed micro waves.

RHEOSTAT
A resistor used for the purpose of controlling current flow.

RECTIFIER
A device for converting alternating current into direct current.

RESISTOR
A device designed to limit the flow of current or to provide a voltage drop.

WATT
The practical unit of electric power. In DC equal to volts times amperes. In AC, true watts are equal to effective volts multiplied by effective amperes, then multiplied by the circuit power factor. In AC circuits, the watt is different from the volt-ampere which is the product of volts times amps.

WAVE GUIDE
A wave guide is an electric conductor consisting of a metal tubing used for the conduction or directional transmission of microwaves.

WAVE LENGTH
Wave length is that distance between corresponding points on two successive AC waves.

DOOR LATCH INTERLOCK TESTS

1. PLACE LOAD IN OVEN CAVITY.
2. Insure that unit is plugged in.
3. Turn oven light on and then off by light switch to insure that oven light is functional.
4. Open door slowly until oven light comes on. The door opening at the top of the door should be less than 5/16 of an inch, or a maximum of 5/16 of an inch, when the light comes on.
5. Left hand interlock test.
   A. Unit unplugged.
   B. Insert an ohmmeter between white lead at the transformer and the neutral side of the line at the terminal board. As the door is closed, meter should show continuity if the switch is in operating condition.

NOTE: *The door latch interlock switch should be checked on every service call.*

GROUP NO. 1 – CONTROL GROUP

Group No. 1 – Switches and Motors.

1. Power Cord.
2. Fan Motor.
3. Stirrer Motor.
4. Door Interlock Switches, 2.
5. Start Switch.
6. Oven Light.
7. Light Switch.
8. Thermal Cut-Out.

POWER CORD
The power cord furnishes power to the unit.

FAN MOTOR
The fan motor is used to cool the magnetron tube and the blower motor.

STIRRER MOTOR
The stirrer motor drives the stirrer blade, which aids in producing a better cooking pattern in the oven.

SECTION 5 AMANA RADARANGE

SERVICE PROCEDURE
COMPONENT DATA

Group 1B — Timing

TIMER MOTOR NO. 1
Controls cooking time in minutes, total time, 30 minutes.

TIMER MOTOR NO. 2
Controls cooking time in seconds, total time, 5 minutes.

TIMER PILOT LIGHT
Furnishes light to the timer which is in operation.

TIMER CONTRACTS

Contact T-30A, Timer No. 1, is a normally open, (N.O.), contact.

Contact T-30B, Timer No. 1, is a normally open, (N.O.), contact.

Contact T-5A, Timer No. 2, is a normally open, (N.O.), contact.

Contact T-5B, Timer No. 2, is a normally open, (N.O.), contact.

All switches are single pole single throw.

DOOR INTERLOCK SWITCHES
The door interlock switches prevent the user from operating the unit without the door closed. There are two (2) switches in the circuit, one on each side of the line to insure positive shutoff when the door is open.

START SWITCH
The unit will not function if the door is not seated against the start switch arm to allow the start switch to function.

OVEN LIGHT
Furnishes light to the oven cavity.

THERMO-CUT-OUT
Prevents the magnetron tube from over-heating.

LIGHT SWITCH
Controls oven light when door is closed.

SERVICE INFORMATION

Effective with the 1970 models, all major household appliances are required to be grounded by Underwriter's Laboratories. Therefore, all units produced after September 2, 1969 will be supplied with a three wire grounded service cord and mechanical grounds to all electrical components, cabinets, liners, and doors, (where required).

WHEN INSTALLING A GROUNDED APPLIANCE IN A HOME THAT DOES NOT HAVE A THREE WIRE GROUNDED RECEPTACLE, UNDER NO CONDITIONS IS THE GROUNDING PRONG TO BE CUT OFF OR REMOVED. IT IS THE PERSONAL RESPONSIBILITY OF THE CUSTOMER TO CONTACT A QUALIFIED ELECTRICIAN AND HAVE A PROPERLY GROUNDED THREE-PRONG WALL RECEPTACLE INSTALLED IN ACCORDANCE WITH THE APPROPRIATE ELECTRICAL CODE.

SHOULD A TWO-PRONG ADAPTER PLUG BE REQUIRED TEMPORARILY, IT IS THE PERSONAL RESPONSIBILITY OF THE CUSTOMER TO HAVE IT REPLACED WITH A PROPERLY GROUNDED THREE-PRONG RECEPTACLE OR THE TWO-PRONG ADAPTER PROPERLY GROUNDED BY A QUALIFIED ELECTRICIAN IN ACCORDANCE WITH THE APPROPRIATE ELECTRICAL CODE.

SERVICING OF GROUNDED PRODUCTS

The standard accepted color coding for ground wires is GREEN or GREEN WITH YELLOW STRIPE. These ground leads are *NOT* to be used as current carrying conductors. Electrical components such as the compressor, condenser fan motor, evaporator fan motor, defrost timer, temperature control, and ice maker (where used) are grounded through the use of an individual wire attached to the electrical component and to another part of the appliance. Ground wires should not be removed from individual components while servicing, unless the component is to be removed and replaced. Should any of the grounded components require servicing which would necessitate removal of the ground wire, IT IS EXTREMELY IMPORT-

SECTION 5 AMANA RADARANGE

SERVICE PROCEDURE COMPONENT DATA

ANT THAT THE SERVICEMAN REPLACE ANY AND ALL GROUNDS PRIOR TO COMPLETION OF HIS SERVICE CALL. UNDER NO CONDITIONS SHOULD A GROUND WIRE BE LEFT OFF CAUSING A POTENTIAL HAZARD TO THE SERVICEMAN AND THE CUSTOMER.

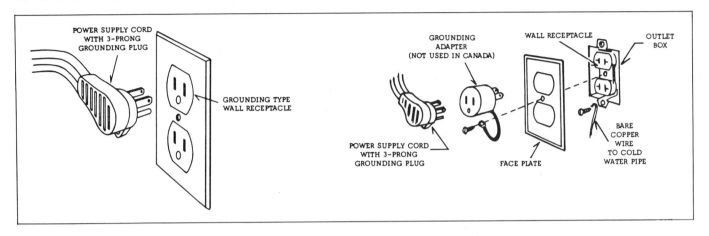

RADARANGE II TIMER OPERATION, *Figure 35 thru 38*

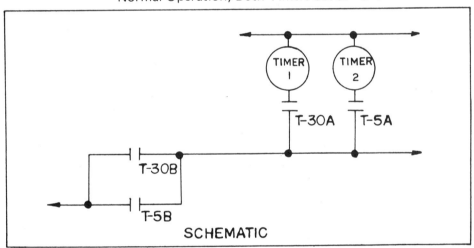

Figure 35

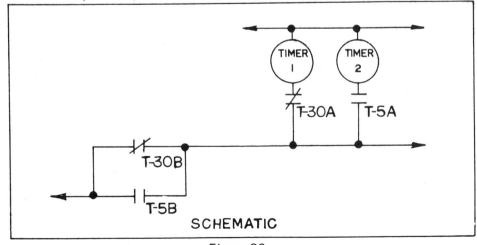

Figure 36

SECTION 5 AMANA RADARANGE

SERVICE PROCEDURE COMPONENT DATA

Normal Operation, Timer No. 1 Set at "O", Timer No. 2 Energized.

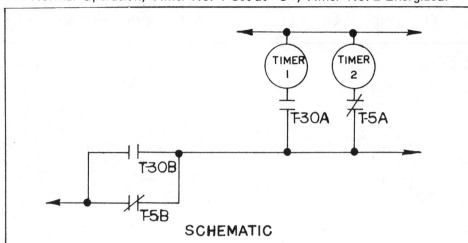

Figure 37

Not Normal Operation, Both Timers Energized.

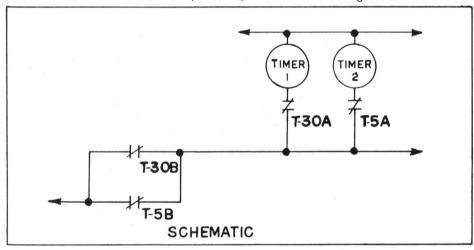

Figure 38

ELECTRONICS GROUP

1. *Magnetron.*
2. *Transformer.*
3. *Bridge.*
4. *Filter Capacitor*

1. MAGNETRON TUBE

The magnetron tube is the heart of the oven.) It converts DC power to RF power.

The magnetron needs a high voltage, (3200 volts) D.C., and a low A.C. voltage, and a magnetic field for normal operation.

The tube consists of a plate, cathode, and filament.

For the tube to operate, electrons must flow through the tube from the cathode to the plate. To get the electrons to flow, high voltage has to exist between the cathode and the plate and the cathode must be heated, *Figure 29.* The heating of the cathode is done by the filament. The filament or heater power is supplied by the three volts A.C. winding on the transformer.

The above description describes the operation of a simple diode. To convert a diode to a magnetron, a magnetic field is applied parallel to the cathode. The magnetic field causes the electrons to spin around the cathode instead of traveling in a straight line from the cathode to the anode.

SECTION 5 AMANA RADARANGE

SERVICE PROCEDURE COMPONENT DATA

The spinning action of the electrons and the configuration of the anode causes R.F. currents to flow on the surface of the anode. An antenna connected to the anode directs the R.F. power flow down the wave guide.

2. TRANSFORMER Figure 39

The transformer used is both a step-up and step-down type and converts the power as follows:

Step-up converts 115 volts A.C. to 3100 volts A.C.

Step-down converts 115 volts A.C. to 3 volts A.C.

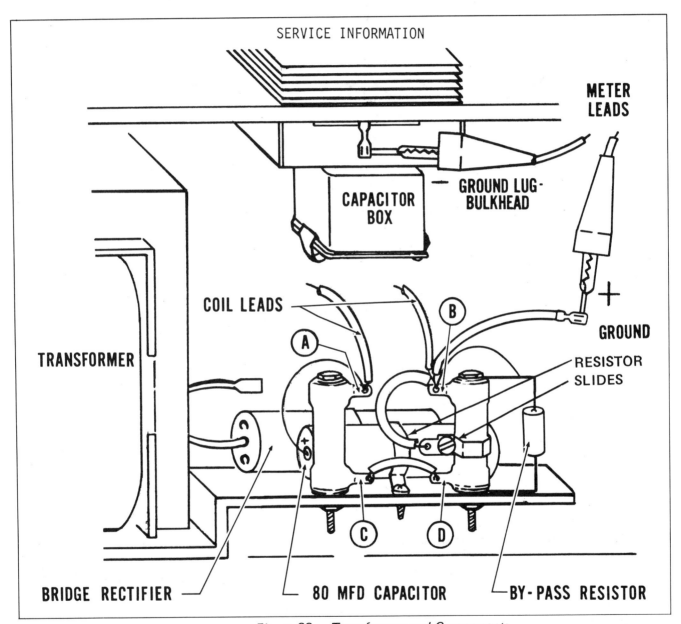

Figure 39 — Transformer and Components

SECTION 5 AMANA RADARANGE

SERVICE PROCEDURE
COMPONENT DATA

BRIDGE RECTIFIER CHECK

1. Set volt ohmmeter to read ohms using X10,000 or higher scale.
2. Check resistance between A and B and A and C. Reverse leads and check again. There should be considerable difference (minimum of 1/2 of scale) between the two readings, approximately a 10 to 1 ratio. If reversing the leads does not change the reading, replace the bridge.
3. Repeat step 2 between D and B and D and C.

NOTE: *Bridge may be checked in unit, but one of the AC wires must be removed.*

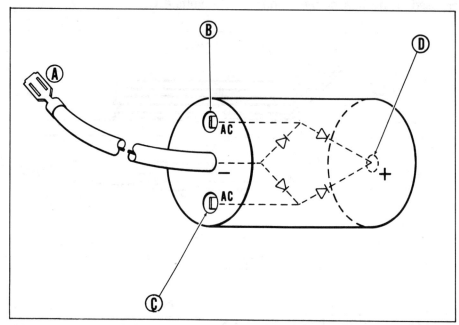

Figure 40 — Rectifier Bridge

CAPACITOR CHECK

1. On those units not having the protective cover, short-out capacitor before testing by placing a screwdriver blade to capacitor cand and then touching capacitor.

2. Remove capacitor can cover and gasket.

3. Remove capacitor protective cover where used: When checking capacitors on those units which have the protective plastic cover on them, slip the side of the cover between the ground clamp and the capacitor while checking the capacitors.

4. Remove both leads from either capacitor. Do not allow the leads to touch ground surfaces.

5. Set ohmmeter on 10,000 ohm scale, check from capacitor to ground. Any reading less than 100,000 ohms indicates a faulty capacitor.

6. After checking or replacing capacitors, dress the tube leads to insure that they are not in such a position as to short to ground.

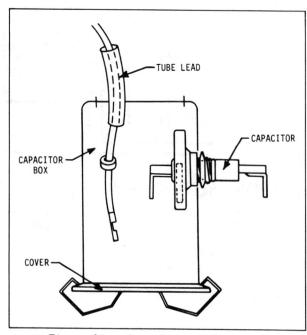

Figure 41 — Capacitor Identification

SECTION 5 AMANA RADARANGE

SERVICE PROCEDURE
COMPONENT DATA

3. *BRIDGE, Figure 40*
   The bridge converts the 3100 volts A.C. from the transformer to 4000 peak volts D.C. which is applied between the cathode and the anode for operation.

4. *FILTER CAPACITOR, Figure 41*
   A filter capacitor by-passes the A.C. current around the magnet coil, thus, allowing ripple free D.C. current to flow in the magnet coil. The end result of this action is a ripple-free, constant magnetic field.

5. *MAGNET COIL*
   Plate current flowing through the magnet coil produces the magnetic field required for magnetron operation, *Figure 42*.

6. By-Pass Current Adjustment Resistor
   The resistors are used to adjust the strength of the magnetic fields supplied to the magnetron. By adjusting the resistor, you regulate the current in the magnetic coil, which affects the magnetic field. The strength of the magnetic field affects the power of the magnetron.

7. By-Pass Capacitors
   Any R.F. current which passes the ferrite rings is picked up by the capacitor and is dissipated through the capacitor to the inside walls of the junction box.

8. Blower Motor
   The cooling blower motor is internally fused so the unit may be used in "built-in" installations. The fuse is embedded in the motor winding. Therefore, if the fuse opens the motor must be replaced.

9. *PROPER HANDLING OF MAGNETRON TUBES*

   A MAGNETRON TUBE, LIKE A RADIO OR TELEVISION TUBE, MUST BE HANDLED WITH A REASONABLE AMOUNT OF CARE. WHEN HANDLING A TUBE, ALWAYS HANDLE BY THE FIN AREA ONLY, AND USE CAUTION NOT TO TOUCH OR STRIKE THE GLASS PORTION AT THE TOP OR BOTTOM.
   THE CARTON USED TO SHIP SERVICE REPLACEMENT TUBES IS OF THE REUSABLE TYPE AND ALL DEFECTIVE TUBES ARE TO BE RETURNED TO THE FACTORY IN THIS CARTON.

RADARANGE MAGNETRON TUBE REPLACEMENT PROCEDURE.

Radarange Model RR-1 and early production of model RR-2 were manufactured with a copper finned magnetron. During early production of the model RR-2, the copper finned tube was discontinued and was replaced with an aluminum finned tube, part No. M3711-1.

Should a copper finned tube require replacement in a model RR-1, it will be necessary to replace the tube, magnetron housing, and the thermal overload. Should tube replacement be required in a model RR-2 containing a copper finned tube, only the tube and thermal overload need to be changed.

The change-out procedures are as follows:

1. Remove the defective tube using normal procedures.

2. If range is an RR-1 with a copper tube, replace the lower magnetron housing support with part No. C35961-1, (If the range is a model RR-2 or a model RR-1 with an aluminum tube already in place, delete this step.)

3. Install part No. M3711-1 aluminum tube using normal procedures.

4. To mount tube thermal cut-out, place cut-out into position against lower tube fin from underside of bulkhead. Engage Clip on bottom fin from right side on top of bulkhead.

5. Connect overload into circuit as shown in sketch "B".

6. Complete assembly using normal procedures.
Summary of parts required for RR-1 and RR-2 conversions:

PART NO.	RR-1 DESCRIPTION	QTY.
M3711-1	Magnetron Tube	1
M3704-2	Thermal Cut-Out	1
M1158-2	Clip Cut-Out	1
C35961-1	Magnetron Housing Support	1
A3176-1	Wire Splice	1
B56219-1	Gasket Washer	2

SECTION 5 AMANA RADARANGE

SERVICE PROCEDURE
COMPONENT DATA

RR-2

PART NO.	DESCRIPTION	QTY.
M3711-1	Magnetron Tube	1
M3704-2	Thermal Cut-Out	1
M1158-2	Clip Cut-Out	1
B56219-1	Gasket Washer	2

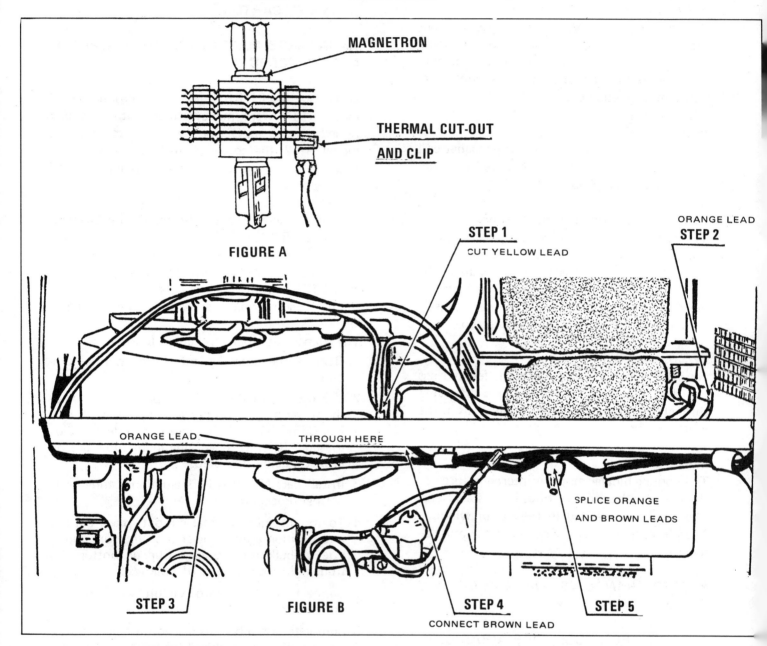

1. Cut the yellow lead ONLY connected to the thermal cut-out. It should be cut at the spade connector and discarded.
2. Remove the orange lead from the thermal cut-out and cut off the terminal.
3. String the orange wire through the bulkhead grommet and along the bottom side of the bulkhead. This will re-route the orange wire.
4. One brown lead from the new thermal cut-out should be connected to the open spade connection on terminal board. The terminal board is mounted to the bottom of the bulkhead.
5. Using a wire nut, splice the orange wire and the remaining brown lead from the thermal cut-out together.

NOTE: *All wires should be secured and held by the wire clamps as illustrated above.*

SECTION 5 AMANA RADARANGE

SERVICE PROCEDURE COMPONENT DATA

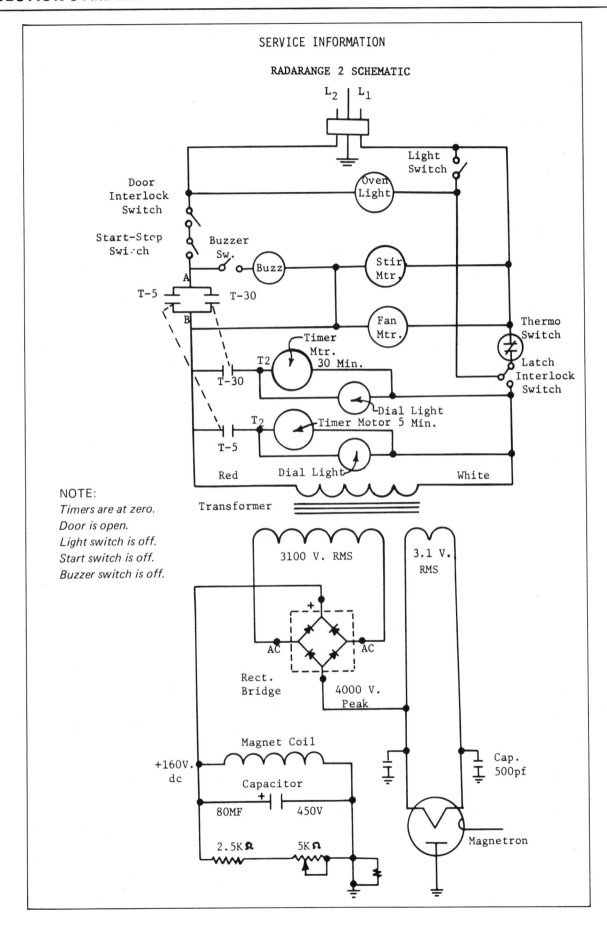

Figure 42

SECTION 5 AMANA RADARANGE

SERVICE PROCEDURE COMPONENT DATA

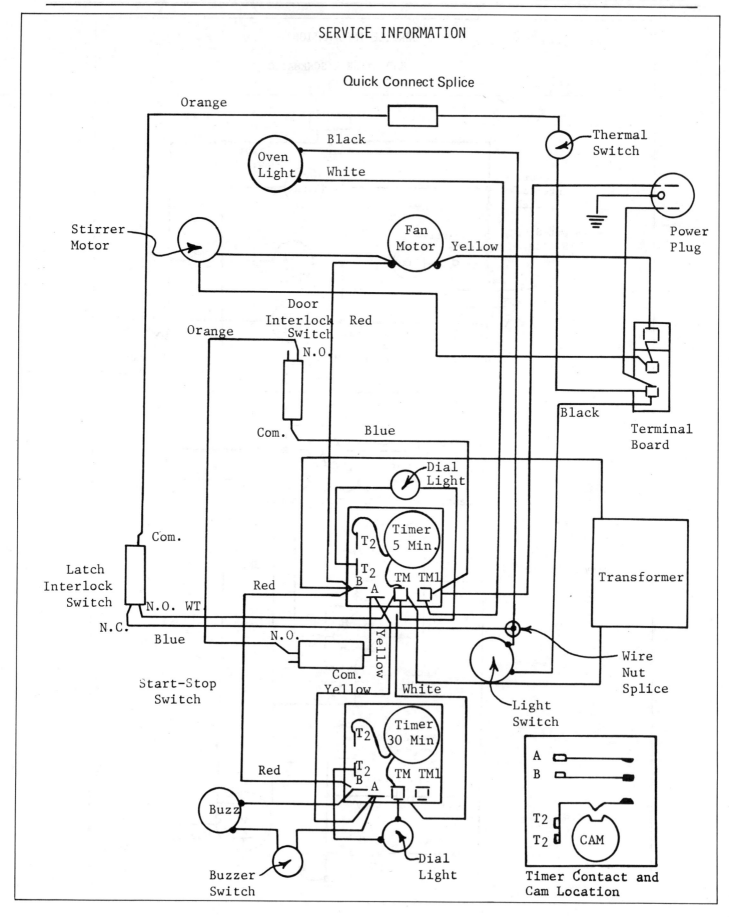

SECTION 5 AMANA RADARANGE

SERVICE PROCEDURE COMPONENT DATA

A - Problems in Transformer and Beyond - Page 79
B - Problems in Control Circuit - Page 80
C - Problems in Open Circuits - Page 81
D - Voltages above 105 V. - Page 82
E - Voltages below 105 V. - Page 83

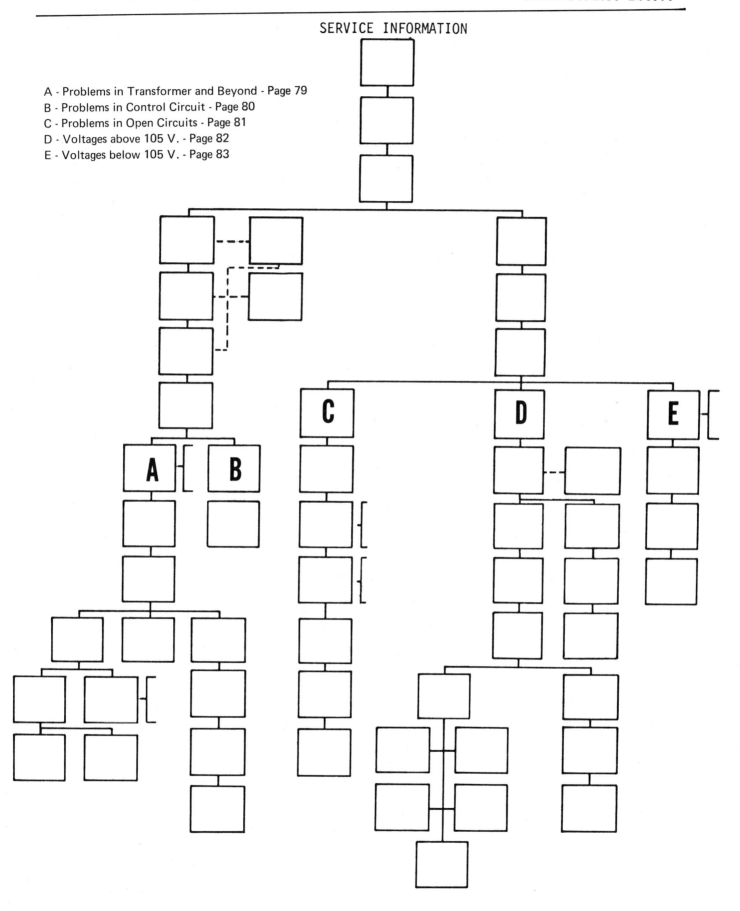

77

SECTION 5 AMANA RADARANGE

SERVICE PROCEDURE
COMPONENT DATA

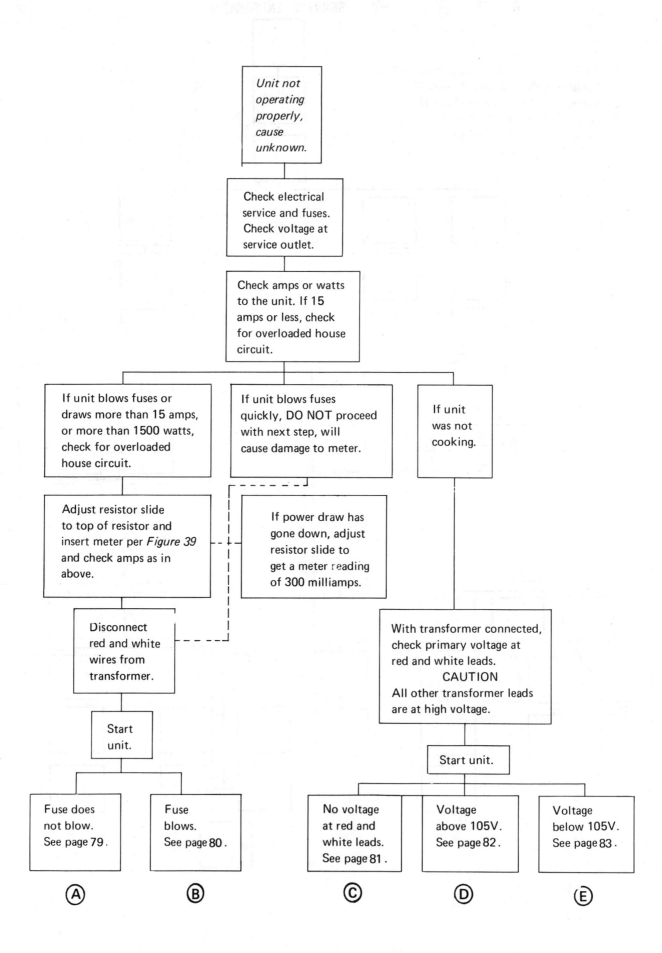

78

SECTION 5 AMANA RADARANGE

SERVICE PROCEDURE COMPONENT DATA

This section pertains to problems in the transformer or beyond.

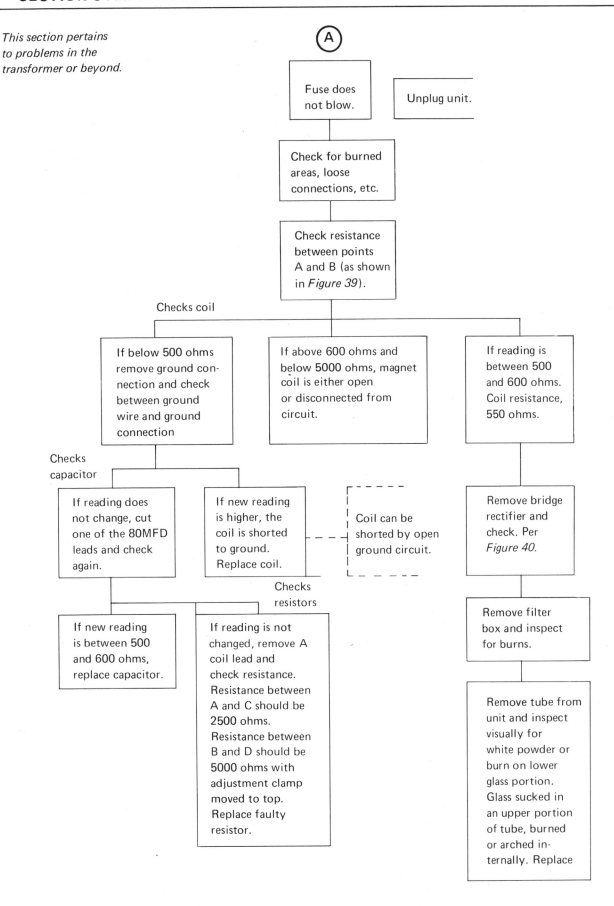

SECTION 5 AMANA RADARANGE

SERVICE PROCEDURE
COMPONENT DATA

This section pertains to checking control circuit for shorts and grounds.

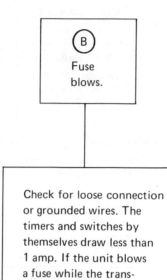

Ⓑ Fuse blows.

Check for loose connection or grounded wires. The timers and switches by themselves draw less than 1 amp. If the unit blows a fuse while the transformer is disconnected, there will be some burned areas visible. Clean and tighten connection.

SECTION 5 AMANA RADARANGE

SERVICE PROCEDURE
COMPONENT DATA

This section pertains to checking for open circuit.

```
┌─────────────────────────┐
│ No voltage at red       │
│ and white leads.        │
└─────────────────────────┘
            │
┌─────────────────────────┐
│ With unit not plugged in.│
│ Close door, set timer,  │
│ push start switch.      │
└─────────────────────────┘
            │
┌─────────────────────────┐
│ Check for continuity. White wire
│ at top of transformer to orange
│ lead at temperature cut-out. No
│ continuity, left door interlock
│ switch defective.
└─────────────────────────┘
            │
┌─────────────────────────┐
│ Check for continuity across
│ temperature cut-out. No
│ continuity, temperature
│ cut-out defective.
└─────────────────────────┘
            │
┌─────────────────────────┐
│ Check continuity at red transformer
│ lead and yellow lead at start switch.
│ No continuity, timer switch. Not
│ making or loose connection in
│ timer area.
└─────────────────────────┘
            │
┌─────────────────────────┐
│ Check continuity across start switch.
│ Check between yellow lead at top
│ timer and orange lead of right hand
│ door interlock switch. No continui-
│ ty, start switch is defective.
└─────────────────────────┘
            │
┌─────────────────────────┐
│ Check continuity between orange
│ lead at start switch and blue lead
│ at top timer terminal. No
│ continuity, right door interlock
│ switch defective.
└─────────────────────────┘
```

NOTE: *Red or white lead must be removed from transformer when taking continuity readings. Also disconnect orange lead to thermo cut-out at Q.C. splice.*

Possible cause —
defective capacitor,
rectifier or tube.
Refer to page 72

SECTION 5 AMANA RADARANGE

SERVICE PROCEDURE
COMPONENT DATA

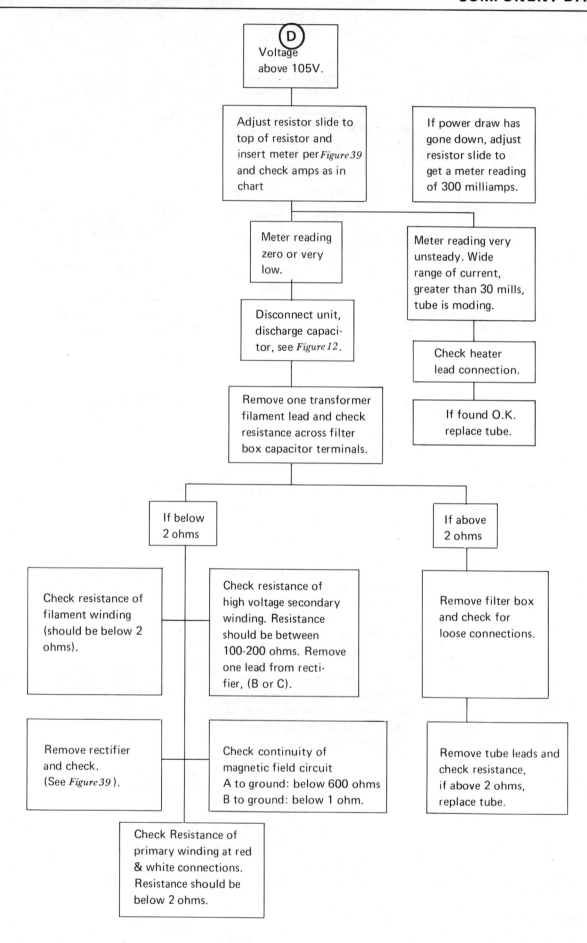

82

SECTION 5 AMANA RADARANGE

SERVICE PROCEDURE
COMPONENT DATA

Voltage below 105V.

If supply voltage is too low for proper operation of the unit, advise the customer to contact a qualified electrical contractor to charge or re-wire the supply.

SECTION 5 AMANA RADARANGE

SERVICE PROCEDURE
COMPONENT DATA

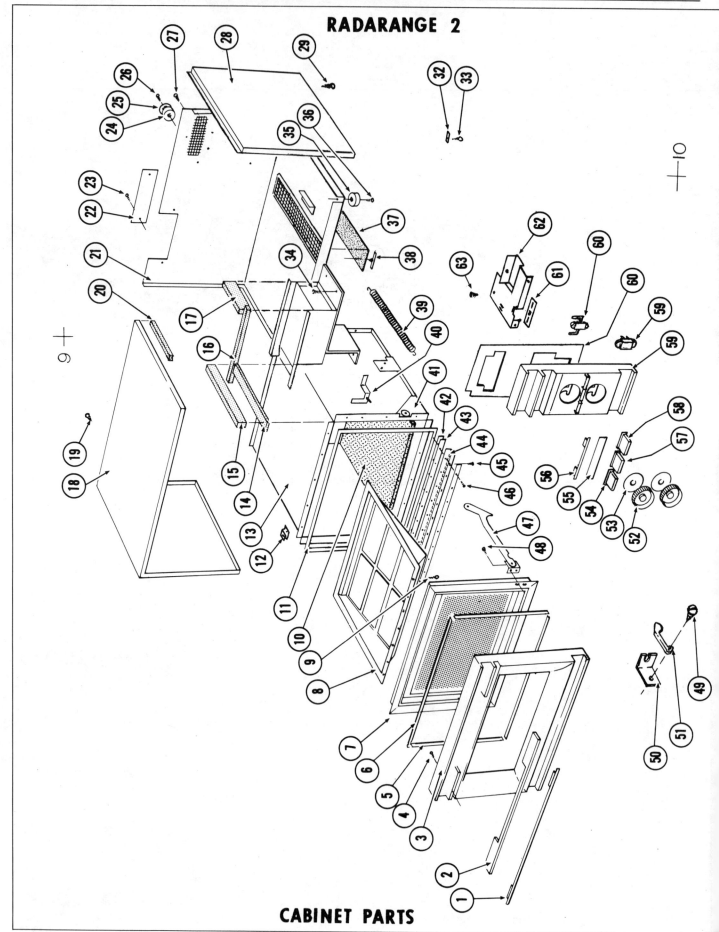

RADARANGE 2

CABINET PARTS

84

SECTION 5 AMANA RADARANGE

SERVICE PROCEDURE COMPONENT DATA

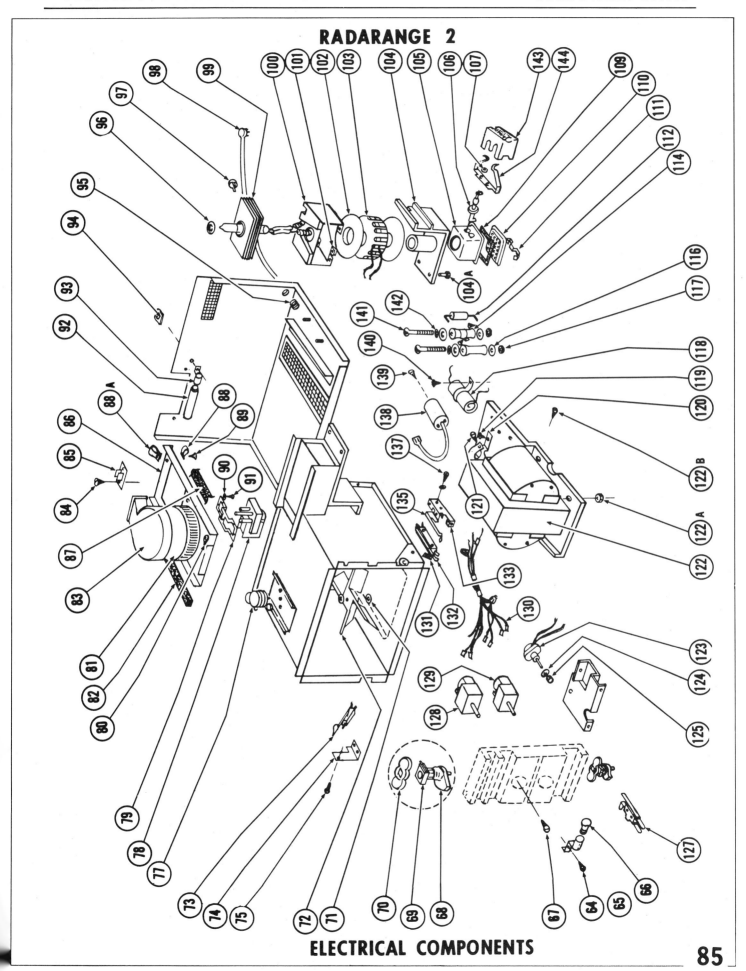

RADARANGE 2

ELECTRICAL COMPONENTS

85

SECTION 5 AMANA RADARANGE

**SERVICE PROCEDURE
COMPONENT DATA**

REF. NO.	DESCRIPTION	RR2
1	Insert, Handle	1
2	Handle, Door	1
3	Door Assembly (Outer)	1
4	Screw, Thread Cutting, Phillips Round Head, #4 X 7/8" Long, Handle Attach	2
5	Gasket, Inner Door Panel, R.H.	1
5	Gasket, Inner Door Panel, L.H.	1
6	Gasket, Inner Door Panel, (Top & Bottom)	2
7	Weldment, Inner Door	1
8	Shield, Light	1
9	Screw, Thread Cutting, Phillips Round Head, #4 X 7/8" Long, (Light Shield)	7
10	Glass Plate	1
11	Gasket, Oven Cavity	1
12	Clip, Cable	2
13	Oven, Weldment	1
14	Insulation Strip, 9" Long, (Oven Weldment)	*
15	Gasket, 9" Long, (Oven Weldment to Outer Case)	*
16	Gasket, 6" Long	*
17	Gasket, 4-1/2" Long	*
18	Outer Case	1
19	Screw, Sheet Metal, #2, Phillips Truss Head, #6-18 X 5/16" Long, (Outer Case to Base Pan)	21
20	Gasket, (6" Long), (Outer Case Flange)	*
21	Chassis, Outer Case Weldment	1
22	Panel, Bulk Access	1
23	Screw, Sheet Metal, #2 Phillips Truss Head, #6-18 X 5/16" Long (Access Panel Attach)	2
24	Rubber Foot	1
25	Washer, Flat, 7/32" I.D. X 7/16" O.D., Foot Attach	1
26	Screw, Thread Cutting, Shoulder Slotted Hex Head #8-32 X 3/4" Long, (Foot Attach)	1
27	Screw, Weldment, 3 Projections for Welding, 3/8" Long	2
28	Access Panel	1
29	Screw, Sheet Metal, #2 Phillips Truss Head, #6-18 X 5/16" Long, (Chassis)	27

SECTION 5 AMANA RADARANGE

**SERVICE PROCEDURE
COMPONENT DATA**

NS	Insulation Batt	AR
NS	Insulation Batt	AR
32	Terminal, Ground, (Bulkhead)	1
33	Screw, Sheet Metal, Slotted Hex Washer Head, #8-18 X 3/8" Long, (Terminal Attach)	4
34	Screw, Machine, Socket Head, #10-32 X 1/2" Long, Magnetron, Housing Attach	2
35	Foot	4
NS	Washer, Foot	4
36	Screw, Thread Cutting, Shoulder Slotted Hex Head, #8-32 X 3/4" Long, (Foot Attach)	4
37	Filter	1
38	Dart Clips	4
39	Spring, Counterbalance	2
40	Reset Arm	1
NS	Button, Reset Arm	1
41	Bracket, Counterbalance, R.H.	1
41	Bracket, Counterbalance, L.H.	1
42	Shim, Door Hinge	1
43	Shim, Door Hinge	1
44	Hinge Door	1
45	Screw, Thread Cutting, Round Head, #1 Phillips, #4-40 X 7/32" Long, (Hinge to Door)	8
46	Screw, Sheet Metal, Flat Head, #1 Phillips, #4-24 X 1/2" Long, (Hinge to Cab)	8
47	Counterbalance Assembly, R.H.	1
47	Counterbalance Assembly, L.H.	1
48	Screw, Thread Cutting, #1 Phillipss, #6-32 X 3/8" Long, (Counterbalance to Door)	4
49	Screw, Shoulder, Phillips Flat Head, (Thread Rolling Tapite) #4-40 X 5/16" Long	1
50	Retainer, Door Latch	1
51	Latch, Door	1
52	Control, Knob	2
NS	Retainer, Control Knob	2
53	Indicator, (Zero to 5 Min.)	1
53	Indicator, (Zero to 30 Min.)	1
54	Start Button	1

SECTION 5 AMANA RADARANGE

SERVICE PROCEDURE
COMPONENT DATA

55	Nameplate, Control Panel Face	1
56	Insert, Escutcheon	1
57	Stop Button	1
58	Light Button	1
59	Control Panel	1
59A	Switch	1
60	Front Panel, Outer Case	1
60A	Buzzer	1
61	Nut, Start and Stop Switch Mounting	1
62	Bracket, Swtich Mounting	
63	Screw, Machine, Truss Head, #1 Phillips, #6-32 X 7/8" Long	1
64	Screw, Thread Cutting, #2 Phillips, #6-32 X 5/16" Long, (Light Socket Mounting)	4
65	Light Socket	2
66	Light Bulb	2
67	Screw, Machine, Phillips Truss Head, #8-32 X 3/8" Long, Timer Mounting	4
68	Cam, Switch	1
69	Bearing Insert, Escutcheon	1
70	Cam, Control Button	1
71	Washer, Clutch	2
72	Stirrer Blade	1
73	Switch, Interlock	1
74	Bracket, Interlock Switch	1
NS	Seal, Access Panel	2
75	Sems Fastener, Machine Screw, Phillips Truss Head, Header Point, #4 X 1/4" Long	2
NS	Speed Nut, Flat Type	2
77	Motor, Stirrer	1
78	Motor, Fan	1
79	Motor Mount, Blower	1
80	"D" Grommet, Bulkhead, Wiring Harness	1
81	Blower Wheel	1
82	Gasket, (Side of BulkheadO	AR
83	Blower, Housing	1

SECTION 5 AMANA RADARANGE

SERVICE PROCEDURE
COMPONENT DATA

84	Screw, Sheet Metal, Slotted Hex Washer Head, #6-20 X 1/2" Long, Terminal Mounting	1
85	Terminal Board	1
86	Bulkhead	1
87	Gasket, (Side of Bulkhead)	AR
88	Clamp, (Power Cord to Bulkhead)	1
88A	Clamp	1
89	Screw, Sheet Metal, Slotted Hex Washer Head, #8-15 X 1/2" Long, (Rectifier to Clamp Mounting)	1
90	Grommet, Motor Mount	4
91	Screw, Shoulder Slotted Hex Washer Head, #6-18 X 5/8" Long, (Mount to Blower Housing)	4
92	Light Bulb (Oven)	1
93	Socket, Oven Light	1
94	Speed Nut, "U" Type, (Oven Light)	1
95	Strain Relief, Power Cord	1
96	Washer, Gasket, Brass Magnetron Tube	2
97	Thermo Cut-Out, Aluminum Magnetron	1
NS	Clip, (Thermal Cut-Out)	1
98	Power Cord	1
99	Magnetron Tube	1
100	Support, Magnetron Housing	1
101	Speed Nut, "U" Type Tinnerman, Magnetron, Support to Housing	4
102	Washer, Nylon, Insulation, (Magnetron Coil)	2
103	Coil, Magnetron	1
104	Weldment, Magnetron Housing	1
104A	Sems Fastener, Thread Cutting, Indented Hex Washer Head, #10-32 X 1/2" Long	4
105	Capacitor Box	1
106	Capacitor	2
107	Washer, Gasket	2
109	Gasket	1
110	Cover, Capacitor Box	1

SECTION 5 AMANA RADARANGE

SERVICE PROCEDURE
COMPONENT DATA

NS	Screw, Sheet Metal, Phillips Truss Head, #6-20 X 3/4" Long	4
111	Bale Strap	1
112	Resistor	1
114	Resistor, Adjustable (5000 OHM 25W)	1
115	Resistor, Fixed, (2500 OHM 25W)	1
116	Washer, Non-Metallic (Resistor)	4
117	Nut, Hexagon	2
118	Capacitor, 80 MFD. 450 D.C. Volts	1
119	Screw, Sheet Metal, Truss Head, #2 Phillips, #6-18 X 3/8' Long	1
120	Clamp, Bridge Rectifier	1
121	Screw, Sheet Metal, Slotted Hex Washer Head, #8-15 X 1/2" Long, (Rectifier Clamp)	1
122	Transformer	1
122A	Nut, Transformer	2
123	Switch, Light	1
124	Washer, Lock, Light Switch	1
125	Nut, Light Switch	1
127	Switch, Start & Stop	1
128	Timer, 5 Minutes	1
129	Timer, 30 Minutes	1
130	Wiring Harness	1
131	Speed Nut, Flat Type	
132	Switch, Latch, Interlock	
133	Bracket, Latch	
135	Spring, Latch	
137	Sems Fastener, Machine Screw, Phillips Truss Head, Header Point, #4-1/4" Long	
138	Bridge Rectifier	1
139	Screw, Machine, Truss Head, () Phillips, #6-32 X 3/16" Long, (Rectifier Connection)	1
140	Screw, Sheet Metal, Hex Washer Head, #6-20 X 3/8" Long, (Capacitor Clamp)	1
141	Screw, Machine, Slotted Round Head, #10-24 X 2-1/2" Long, (Resistor Attach)	2
142	Washer, Flat, 13/64" I.D. X 1/2" O.D. X 1/32" Thick, (Resistor)	2
143	Shield, H.V. Terminal	1
144	Clip, H.V. Terminal	1

SECTION 6 FRIGIDAIRE MICROWAVE OVENS

SERVICE PROCEDURE COMPONENT DATA

CAPACITOR REPLACEMENT — MUST WATCH POLARITY:

When the 80 mfd. capacitor in the electro-magnetic magnetron circuit is replaced, it is necessary that the positive capacitor lead be connected to the positive side of the circuit and the negative lead to the negative side.

Some replacement capacitors have the mounting bracket reversed, as compared to the original capacitor, and if the serviceman is not watching polarity, he can easily connect it improperly.

If the capacitor polarity is wrong, it will act as a low resistance shunt around the magnetic coil and the magnetic field on the magnetron will be very weak. The result will be a high magnetron current, but little or no microwave power will be delivered to the oven.

CAPACITOR, RFI, MICROWAVE OVEN — NEW

The radio frequency interference capacitor assemblies for the microwave oven have been replaced by single capacitors of 9400 mfd capacity. The new capacitors are shown assembled on the capacitor box, *Figure A*.

The same part number, is used for the new capacitors as was used for the previous capacitor assemblies.

When one of these capacitors has shorted out, be sure to check the rectifier, the relay and the 80 mfd. filter capacitor, as they might have been damaged when the RFI capacitor shorted.

DOORS, MICROWAVE OVEN — PINNED TOGETHER

The eight cap head screws used to mount the outer door panel to the inner door panel and choke assembly, have been replaced by seven roll pins. Pinning the door together eliminates the possibility of a gap between the two panels. Some customers might find this gap objectionable from an appearance standpoint, particularly along the top edge where the aluminum die casting can be seen between the outer panel and the secondary seal.

The locating dimensions of these pins are shown in *Figure B*. They may be removed, if necessary, by driving them through. When replacing one of the panels, the serviceman may use either the pins or screws.

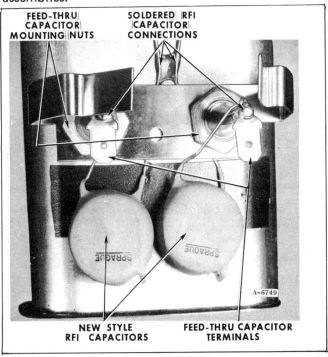

Figure A

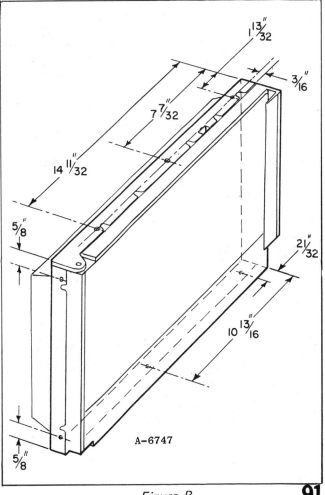

Figure B

SECTION 6 FRIGIDAIRE MICROWAVE OVENS

SERVICE PROCEDURE COMPONENT DATA

One pin may be installed at the 7-7/32 dimension to eliminate a gap on a door assembled with screws.

Holes must be drilled for the pins using a #40 drill at the dimensions shown. Squeeze the panels together while drilling to eliminate the possibility of a gap. After drilling, take the two panels apart to remove any plastic and metal chips.

NOTE: *This gap is an appearance item only and has no microwave leakage implications whatsoever.*

NEW PM MAGNETRON, NEW POWER SUPPLY IN MICROWAVE OVENS

The microwave oven now in production has a new magnetron equipped with built-in permanent magnets and, therefore, does not need a magnetic coil to supply its magnetic field. The high voltage power supply also has been changed. A 2,000 volt transformer and a voltage doubler circuit are now used to supply the 4,000 volts D.C. for the magnetron.

The new magnetron is shown in *Figure C*. Its construction is shown in the cutaway drawing in *Figure D*. As shown in the figure, the magnets are four ceramic rings surrounding the lower portion of the magnetron. The magnetron housing, as well as the magnets, is now an integral part of the magnetron tube assembly. Thus, everything in *Figure D,* except the waveguide, is included in the magnetron assembly.

The new circuit is shown in *Figure E*. As shown, the RFI capacitors are no longer needed and are now omitted.

THERMAL FUSE PROTECTS AGAINST HIGH CURRENT

The start switch, interlock switches, timer and relay have not been changed. As shown in the schematic, a thermal fuse is now connected in series with the transformer primary. This fuse is contained inside a 10-ohm resistor connected in series with the high voltage secondary winding of the transformer.

If the secondary current is too high because of a short or for any other reason, the 10-ohm resistor will heat and cause the thermal fuse to melt, opening the primary circuit of the transformer. The fuse opens when it reaches a temperature of 330°F.

If the fuse opens, it is important to check all components of the high voltage circuit for shorts. The rectifier especially should be checked.

The resistor-fuse board is shown in place in *Figure F* and removed in *Figure G*.

CAPACITOR AND RECTIFIER FOR VOLTAGE DOUBLING

The operation of the voltage doubler power supply circuit is as follows:

Figure C

SECTION 6 FRIGIDAIRE MICROWAVE OVENS

SERVICE PROCEDURE COMPONENT DATA

When the voltage in the transformer high voltage secondary winding is negative on the grounded end, *Figure H*, electron flow through the rectifier will be in the direction shown by the arrow and will charge the capacitor to 2,000 volts with the polarity shown.

On the next half of the cycle, when the secondary voltage is positive on the grounded end, the rectifier will not conduct and, therefore, can be considered an open in the circuit. See *Figure I*. The capacitor is still charged and the polarity of the voltage on the transformer now is added directly to the capacitor voltage, giving 4,000 volts total which is applied between the cathode and anode of the magnetron.

The magnetron voltage, therefore, is that shown in *Figure J*. It will be 2,000 volts on one half the A.C. cycle and rise to 4,000 volts on the other half cycle.

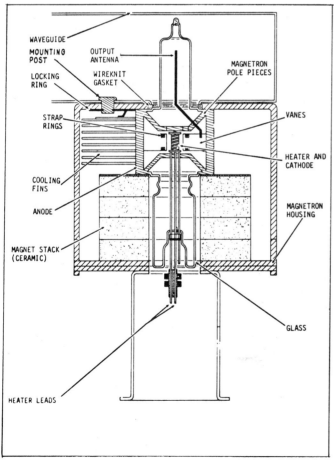

Figure D

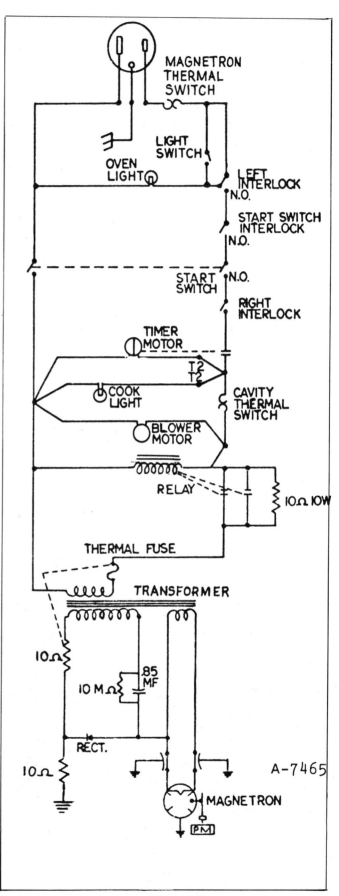

Figure E

SECTION 6 FRIGIDAIRE MICROWAVE OVENS

SERVICE PROCEDURE COMPONENT DATA

IMPORTANT NOTE: *Since the capacitor can retain its charge for some time it is important to DISCHARGE THE CAPACITOR before making tests or replacements of any components in the high voltage circuit. If this is not done, the technician could receive a serious electrical shock. The capacitor can be discharged by placing a screwdriver with an insulated handle across its terminals or between one of the feed-through capacitor terminals and ground, Figure 23A.*

RECTIFIER MUST WITHSTAND 4,000 VOLTS

As shown in *Figure I,* the diode rectifier will have 4,000 volts applied to it on one half of every A.C. cycle. For this reason, the rectifier should be carefully checked whenever a thermal fuse is blown.

To check the rectifier, discharge capacitors and disconnect one lead and connect an ohmmeter across the rectifier terminals. The ohmmeter should be set on a high resistance scale. Measure the resistance and then reverse the ohmmeter leads and measure the resistance again. One of the resistance measurements should be at least ten times as high as the other. If it is not, the rectifier should be replaced.

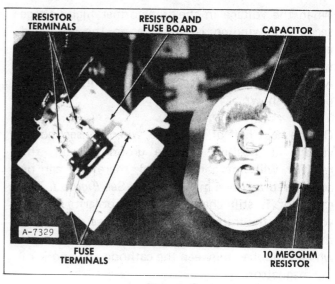

Figure G

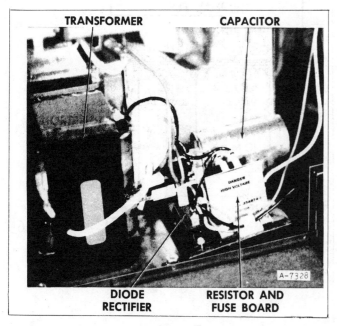

Figure F

When a new rectifier is installed the arrow symbol must point toward the rear.

ALLOWING FOR CAPACITOR TOLERANCE

The .85 mfd. oil-filled capacitor used in the voltage doubler circuit has a 6% tolerance in capacitance. Each capacitor is color coded, when manufactured, with a dot of paint on the side 1/4" from the terminal end. A black dot indicates a tolerance of −6% to −3%; a white dot, −3% to 0%; a yellow dot, 0% to +3%, and a red dot, +3% to +6%.

To compensate for the tolerance in capacitance, two taps are provided on the high voltage winding of the transformer. Thus, to compensate for a capacitance on the low side of the tolerance, a slightly higher voltage from the transformer is used.

If it is necessary to replace the capacitor, be sure to check the color of the dot on the side of the new capacitor. If it is a black or white dot the capacitor lead to the transformer should be connected to the outside tap (4) on the transformer. If it is a red or yellow dot, the connection should be to the inside tap (4A).

When a capacitor is replaced, be certain that the 10-megohm resistor is not located within one-half inch of any grounded part.

SECTION 6 FRIGIDAIRE MICROWAVE OVENS

SERVICE PROCEDURE COMPONENT DATA

REPLACING A MAGNETRON

Magnetron failures with the new tube will be indicated by the same signs as with the previously used tube. A great decrease in power delivered to the oven (indicated by longer required cooking time, as much as twice as long) usually indicates a magnetron that is "moding," unstable in frequency. A tube with this characteristic must be replaced.

TO REPLACE A MAGNETRON

1. Discharge the capacitor.
2. Remove the heater leads from the feed-through capacitor terminals.
3. Release the magnetron by pushing in with one hand on the release lever, *Figure C.* Hold the magnetron assembly with the other hand to prevent it from falling. Pushing on the release lever rotates the locking ring, *Figure D,* and releases the magnetron from the mounting posts.
4. Lower the tube assembly carefully and remove from the oven.
5. If the feed-through capacitors are in good condition they may be removed and installed on the replacement magnetron.
6. Check the ferrite rings on the magnetron leads to insure that neither rings nor leads are near any grounded parts.
7. To install the new magnetron, insert the antenna glass carefully into the waveguide. Hold the tube assembly firmly against the waveguide with one hand and pull the release lever to the rear with the other.
8. Check to insure that the magnetron assembly is properly locked in place on all four mounting posts. Improper installation will result in improper functioning of the magnetron.

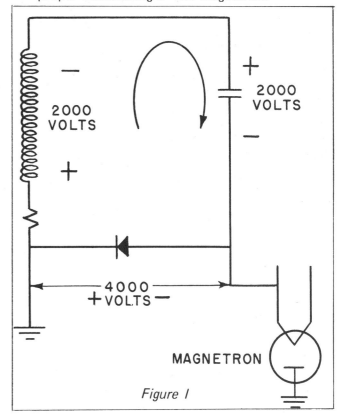

Figure I

SWITCH, THERMAL — MAGNETRON

The magnetron thermal switch was removed for *Figure C* to allow the magnetron release lever to be seen better. This thermal switch is mounted on the upper rear edge of the oven liner, as close to the magnetron as possible. See *Figure K.*

The part number for this switch is the same as that used with the electromagnetic magnetron circuit.

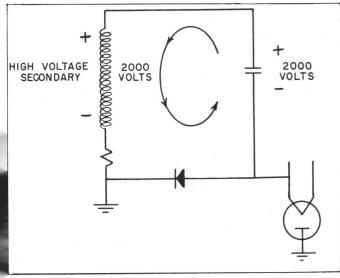

Figure H

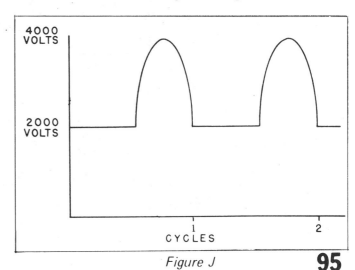

Figure J

SECTION 6 FRIGIDAIRE MICROWAVE OVENS

SERVICE PROCEDURE COMPONENT DATA

OTHER PARTS

Testing and replacing other parts of the microwave oven is the same as on previously produced units. The new wiring diagram is shown in *Figure K*.

NEW PART NUMBERS

The new part numbers involved are as follows:

9948226	Magnetron Tube
9948227	Transformer
9948228	Resistor — 10 Megohm
9948229	Capacitor — .85 MFD.
9948230	Capacitor Strap
9948231	Diode Rectifier
9948813	Fuse, with Sleeve
9948814	Resistor, Heating, 10 ohms, 8 watts
9948815	Resistor, Ground, 10 ohms, 5 watts
9948816	Terminal Board
9948233	Oven Liner
9948822	Baffle, Air
9948823	Clip, Baffle
9948824	Gasket, Magnetron

All microwave ovens with serial numbers starting with a 3 have these parts.

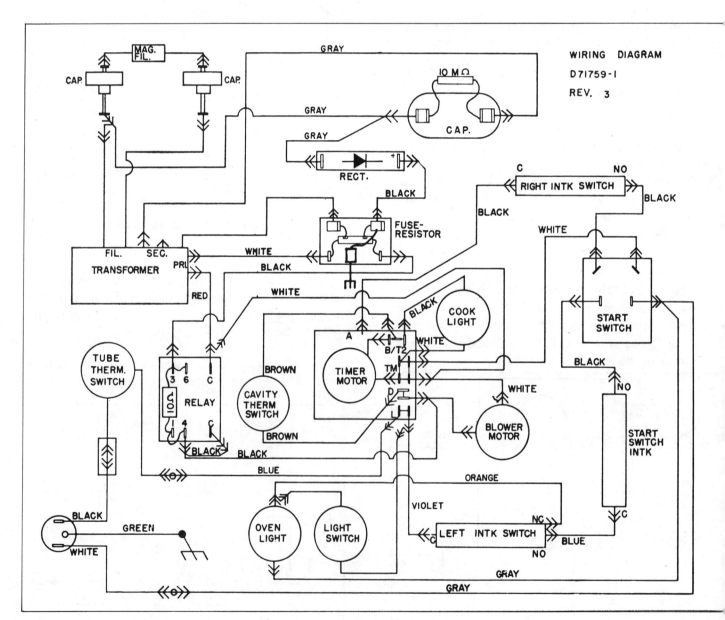

Figure K

SECTION 6 FRIGIDAIRE MICROWAVE OVENS

SERVICE PROCEDURE COMPONENT DATA

The ground resistor is a new part added for the serviceman's protection when checking the magnetron current. It is suggested this part be added when servicing a unit which does not have it.

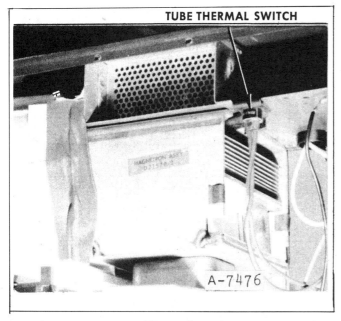

Figure L

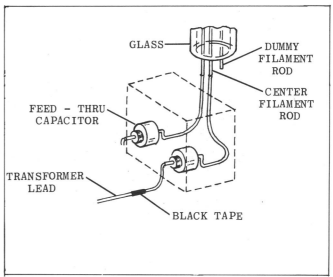

Figure M

TRANSFORMER, CORRECT PHASING NECESSARY WITH NEW CIRCUIT

With the new high-voltage circuit for the microwave oven, it is important that the transformer leads be connected in the proper phase relationship. If they are not connected correctly, the diode rectifier may fail prematurely.

The transformer filament lead marked with black tape must be attached to the right-hand feed-thru capacitor. The inner terminal of this capacitor must be connected to the lead from the center filament rod of the magnetron, as shown in *Figure M*. The other transformer filament lead and the lead to the outer filament rod in the magnetron are connected to the left-hand feed-thru capacitor.

If the transformer does not have a lead marked with black tape, the phasing can be checked with a DC voltmeter and a "D" size flashlight battery. The voltmeter is set on a 25 to 60 volt D.C. scale and connected to transformer lead No. 3 and terminal No. 4 as shown in *Figure U*. Lead numbers 5 and 6 are touched to the negative and positive terminals of the battery as shown in the figure. To prevent discharging the battery, a momentary connection only should be made.

With the connections as shown, if the meter needle first deflects *negative* when the connection is *made*, and *positive* when the connection is *broken*, place a piece of black tape on lead No. 5 within one inch of the terminal. If the meter needle deflects *positive* when the connection is *made* and *negative* when it is *broken*, place the black tape on lead No. 6.

Connect the taped lead to the right-hand feed-thru capacitor as described above.

ANTENNA ORIENTATION— NOT IMPORTANT IN NEW CIRCUIT

With the permanent magnet circuit, the magnetron antenna may be aimed in any direction in the waveguide. Microwaves are emitted from the antenna in a 360° pattern so the wave propagation is not affected by the antenna angle.

The magnetron used with the electromagnetic circuit in the first microwave ovens has its fins attached in such a manner that, if the magnetron is turned 90°, an abnormal stress is placed on the glass, possibly causing a tube failure. The easiest way to correctly position the old stype magnetron is to look at the antenna and see that the flat side faces down the waveguide.

SECTION 6 FRIGIDAIRE MICROWAVE OVENS

SERVICE PROCEDURE COMPONENT DATA

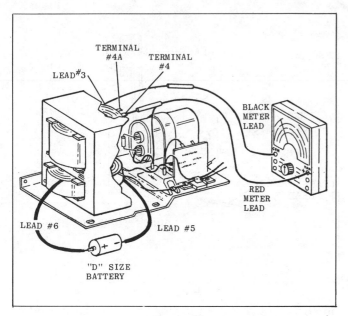

Figure N

The microwave power which is not delivered to the oven is reflected back down the waveguide and develops additional heat in the magnetron. This can shorten the life of the magnetron, *Figure O*.

For these reasons, the stirrer should be checked whenever the complaint is slow heating.

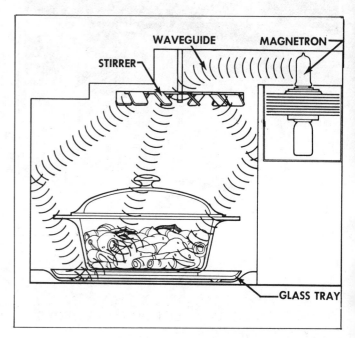

Figure O

STIRRER INVERTED WILL CAUSE LOW POWER, LONGER HEATING TIME

A few cases have been reported of microwave ovens delivered with the stirrer mounted upside down. With this condition, the microwave power in the oven is reduced by one-third to one-half of normal. The heating or cooking time is increased correspondingly.

SECTION 7 GENERAL ELECTRIC MICROWAVE OVENS

SERVICE PROCEDURE COMPONENT DATA

GENERAL ELECTRIC COUNTERTOP MICROWAVE OVEN

PRECAUTIONS FOR PROPER AND SAFE SERVICE

The microwave oven has been designed, built and tested to rigid industry standards. If damaged or serviced in a way that will allow microwave energy to escape, the microwave oven might be hazardous. It is important that you read and observe the following precautions in servicing the oven:

1. DO NOT ATTEMPT TO OPERATE THE MICROWAVE OVEN WITH DOOR OPEN. The microwave oven has been provided with several interlocks to protect you. Do not attempt to "fool" them. They are intended to make sure the door cannot be opened while microwave power is on.

2. DO NOT ATTEMPT TO OPERATE THE MICROWAVE OVEN WHEN THE MAGNETRON OR WAVEGUIDE IS LOOSE OR DISASSEMBLED.

3. DO NOT OPERATE THE MICROWAVE OVEN UNTIL AFTER REPAIR, IF ANY OF THE FOLLOWING CONDITIONS EXIST:

   A. Door does not close firmly against front frame.

   B. Broken hinge.

   C. Door seal damaged.

   D. Bent or warped door.

   E. Any other visible damage to oven.

4. DO NOT OPERATE THE MICROWAVE OVEN UNTIL AFTER REPLACEMENT OF ANY DEFECTIVE PARTS IN THE *INTERLOCK, OVEN DOOR, MAGNETRON,* AND *WAVEGUIDE.*

5. DO NOT LEAN OR PLACE HEAVY SERVICE TOOLS OR EQUIPMENT ON OVEN DOOR.

SAFETY CHECKS

In the interest of good service, and for maximum safety to the user, the following safety checks should be performed upon the conclusion of any service to this appliance:

1. Check electric cord and plug. Advise user if improperly grounded wall receptacle is being used, such as a non-grounded adapter.

2. Check cord grounding lead to frame. Check for connection and continuity between ground terminal on plug and oven frame.

3. Check plastic stirrer cover for proper installation inside oven. Cover should rest in bracket at rear and be fastened at front with screws.

4. Make physical check of door for build-up of soil on door seals or any possible damage. Check door fit and adjustment according to instructions.

HOW TO USE THE MICROWAVE OVEN

A. Be certain glass cooking shelf (tray) is in place at bottom of oven.

B. Place food in oven. Do not use metal utensils.

C. Close door.

D. Set controls:

   1. Push on-off switch. Sound of blower noise should be heard, *Figure 43*.

Figure 43

SECTION 7 GENERAL ELECTRIC OVENS

SERVICE PROCEDURE COMPONENT DATA

2. Set timer for desired cooking time, *Figure 44*. Wait for oven light to come "on" (approximately seven seconds delay).

Figure 44

3. When oven light is "on", push cook switch — Cook light should come "on" immediately, indicating that microwave power is "on", *Figure 45*.

Figure 45

E. When the preset cooking time has elapsed, a bell will sound and microwave power will automatically be turned "off".

NOTE: *The cooking operation can be stopped at any time by opening the door, turning the timer to "O", or pushing "off" switch. To re-start, set controls as above.*

CIRCUIT OPERATION

The description and operation of the microwave oven circuit is as follows (refer to oven schematic):

When the ON-OFF SWITCH is pushed to "on", the switch contacts close and complete a circuit to the BLOWER MOTOR, MAGNETRON FILAMENT (HEATER) TRANSFORMER, and the TIME DELAY RELAY COIL *Figure 46*. After about seven seconds, the time delay relay operates and its contacts close. (The time delay is necessary to allow the magnetron filament to warm up before applying high voltage.)

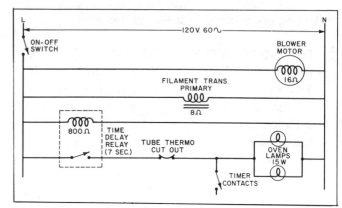

Figure 46 — Schematic, Thermal Cut-Out

When the time delay relay contacts close, it completes a circuit through the magnetron THERMAL PROTECTOR (N.C.), and turns on the OVEN LIGHTS. *This is the indicator to the user that the magnetron is warmed up and ready to use.*

When the timer is set, the contacts close and complete a circuit through a set of contacts in one-half of the dual LATCH SWITCH and up to the COOK SWITCH. With the oven lights "on", *when the COOK SWITCH is pushed to "on",* it completes two circuits:

1. One circuit is through both sets of contacts in the COOK SWITCH which establishes a circuit for, and energizes, the COOK RELAY, TIMER MOTOR and STIRRER MOTOR through the other half of the LATCH SWITCH, *Figure 47*.

2. The second circuit energized is the primary winding of the HIGH VOLTAGE TRANSFORMER, through the three OHM (3Ω) SURGE LIMITING RESISTOR and the DOOR SAFETY SWITCH. This also turns "on" the COOK LAMP which indicates that the high voltage is "on" and the food should be cooking.

SECTION 7 GENERAL ELECTRIC OVENS

SERVICE PROCEDURE COMPONENT DATA

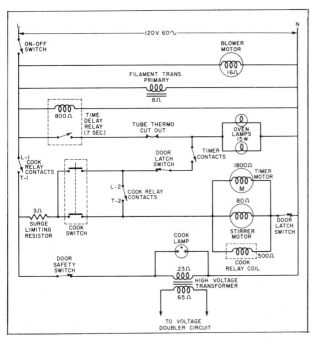

Figure 47 — Schematic, Cook Switch

NOTE: *Just as soon as the COOK RELAY is energized it closes two (2) sets of contacts which "latch in" the above two circuits. This latching effect is necessary to maintain the circuits since the COOK SWITCH is only a momentary switch.*

The high voltage transformer secondary winding provides approximately 2300 volts A.C. to the VOLTAGE DOUBLER CIRCUIT.

VOLTAGE DOUBLER CIRCUIT

The magnetron requires a high D.C. voltage source for operation. This voltage is supplied by a half-wave voltage doubler circuit consisting of a HIGH VOLTAGE CAPACITOR and a DIODE BOARD.

The 2300 volt secondary winding of the transformer is connected to the *capacitor* and *diodes*. The capacitor increases the voltage to about 3800 volts due to its peak voltage charging capabilities. The diodes rectify the voltage, and the resultant half-wave D.C. voltage is applied to the magnetron. The negative (-) output from the D.C. circuit is connected to the filament of the magnetron. The positive (+) output is connected to the plate of the magnetron which is *grounded* to the chassis.

The diodes are therefore connected with *opposite polarity* from the magnetron in order that the *capacitor will charge through the diodes* during one-half cycle, *Figure 48*, and will *discharge through the magnetron* on the alternate half cycle, *Figure 49*. On the discharge cycle, the capacitor

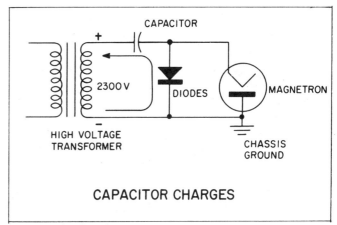

Figure 48 — Capacitor Charge to Peak Voltage

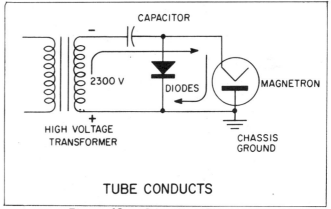

Figure 49 — Capacitor Charge to Aid Transformer

aids the transformer voltage, thereby providing the voltage doubling action.

It should be pointed out that while the circuit is called a *"voltage doubler"*, it does not actually provide *double* the applied voltage.

OUTER CASE

1. Disconnect power.

2. Remove screws at sides and rear of case.

SECTION 7 GENERAL ELECTRIC OVENS

SERVICE PROCEDURE COMPONENT DATA

3. Push outer case slightly to rear to free retaining clips along front edge. Lift off case.

4. DISCHARGE HIGH VOLTAGE CAPACITOR BY SHORTING ACROSS ITS TERMINALS. TWO PLASTIC HANDLE SCREWDRIVERS CAN BE USED. SIMULTANEOUSLY TOUCH EACH CAPACITOR TERMINAL AND THE SCREWDRIVERS TOGETHER, *Figure 50*.

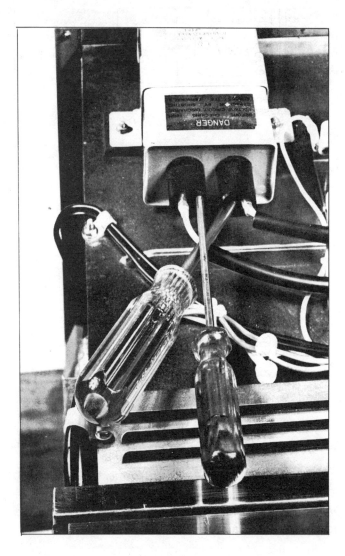

Figure 50 — Discharging Capacitor

CONTROL PANEL

The control panel contains the operational controls (on-off switch, timer and cook switch) and can be removed as an assembly for service:

1. Disconnect oven.

2. Discharge high voltage capacitor.

3. Disconnect all leads from controls.

4. Remove plastic trim at bottom of control panel. The trim is fastened with one screw.

5. Remove the two panel screws exposed after a trim is removed, *Figure 51*.

Figure 51 — Control Panel Removal

6. Slide control panel down slightly, to free clips at top, and lift panel out.

7. Remove instrument panel from trim by removing two screws at the top and two screws at bottom, *Figure 52*. Remove timer knob.

ON-OFF SWITCH

The main on-off switch is a pushbutton-operated, 12 amp, 125 volt miniature switch mounted to a spring-loaded actuating trigger assembly, located on the control panel, *Figure 52*.

The switch and trigger mechanism is mounted to the control panel from the front by two screws. The control panel must be removed to service the switch and actuator assembly.

When the switch is turned "ON", it energizes the blower, filament transformer and time delay relay.

TIMER

The timer is a 25 minute timer assembly consisting of the motor, contacts, and bell located on the

SECTION 7 GENERAL ELECTRIC OVENS

SERVICE PROCEDURE COMPONENT DATA

control panel. The timer is mounted from the front by three screws. The control panel must be removed to service the timer, *Figure 52*.

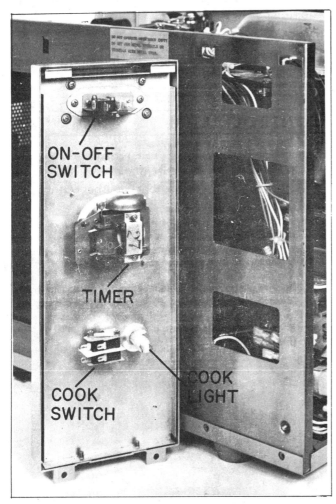

Figure 52 – On-Off and Cook Switch

The normally open (N.O.) switch is an 8 amp, 125 volt, pushbutton-type miniature switch mounted to the side of the timer and actuated or closed when the timer is set. This switch, along with the cook relay contacts, controls the circuit to the *timer motor, stirrer motor,* and *cook relay coil.* When the timer returns to "O" position, the timer contacts open and a mechanical bell chimes once.

The timer can control cooking times from 0 to 25 minutes. The dial is marked in one(1)-minute intervals over the full 25 minute span. In addition, the first five minutes is sub-divided into 15-second intervals.

COOK SWITCH

The cook switch consists of two (2) 12 amp, 125 volt miniature momentary-type pushbutton switches. They are mounted together, side-by-side, and actuated by a pushbutton operated bar to make the assembly serve as a double pole switch, *Figure 52*.

The switch is mounted to the control panel by a lock nut, and is serviced as an assembly.

The cook switch serves two functions:

1. Momentarily establishes a circuit to energize the cook relay coil, which provides a latching circuit around the cook switch.

2. Momentarily establishes a circuit to the high voltage transformer through a 3 OHM surge (current limiting) resistor. The transformer circuit is then maintained by a set of contacts in the cook relay.

The cook switch can be checked by a continuity test. The contacts should be *closed* when the button is "pushed" and it should *open* when the button is released.

COOK RELAY

The cook relay is a double pole type with a 120 volt coil and two sets of 15 amp contacts. The relay is mounted to a mounting plate by two screws and is located on the top of the oven, *Figure 55*.

The cook relay serves two functions:

1. One set of contacts (L2-T2) maintains the circuit for the relay coil. This latching effect is required, since the initial circuit is established by the momentary cook switch.

2. The other set of contacts (L1-T1) provides the circuit for the high voltage transformer.

NOTE: *The contact numbers do not appear on the relay – for circuit reference only.*

The relay coil is energized when all controls are set and the cook switch is "pressed"

SECTION 7 GENERAL ELECTRIC OVENS

SERVICE PROCEDURE COMPONENT DATA

HOW TO TEST

1. Unplug oven and remove outer case.

2. Discharge high voltage capacitor.

3. Check continuity of relay coil. The *normal* resistance should be approximately 500 OHMS.

4. Operate relay manually and check continuity of each set of contacts.

OVEN LIGHTS

Two 15 watt, 120 volt incandescent lamps with bayonet-type bases are used for oven cavity illumination, *Figure 53*.

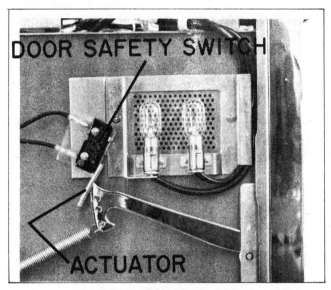

Figure 53 — Oven Lights

The lamps are accessible through a small panel on the left side of the oven case.

During normal operation, the lamps are also used as an indicator that the oven is ready for use. The lamps come "on" after the seven-second time delay relay closes.

TIME DELAY RELAY

A seven-second time delay relay is used to provide a *warm-up* period for the magnetron filament before high voltage is applied to the tube, *Figure 54*.

The relay consists of a 120 volt coil which operates as soon as the "on-off" switch is pressed. Approximately seven seconds after the coil is energized, the relay contacts close. This completes a circuit up to the timer switch.

The time delay relay is a hydraulic type which produces its "time delay" by the time it requires the magnetic field to lift a metal slug through a fluid filled sealed chamber. When the slug reaches the top of the chamber, the concentration of the magnetic pull closes the switch contacts.

The relay is position-sensitive and must be mounted with the sealed chamber at the bottom, as shown in *Figure 54*.

Figure 54 — Time Delay Relay

HOW TO TEST

1. Unplug oven and remove outer case.

2. Discharge high voltage capacitor.

3. Remove leads from one side of relay coil and check continuity. A normal resistance is approximately 800 OHMS.

SECTION 7 GENERAL ELECTRIC OVENS

SERVICE PROCEDURE COMPONENT DATA

4. Connect leads and plug in oven and turn "ON". If coil is good and voltage available to coil, *normally*, the contacts should close in seven seconds and turn "ON" the oven lights.

HOW TO REPLACE

1. Unplug oven and remove outer case.

2. Discharge high voltage capacitor.

3. Remove two screws mounting relay board to oven cavity, *Figure 55*.

4. Lift relay board and remove screws holding time delay relay.

5. Unsolder necessary leads and transfer to new relay.

Figure 55 – Relay Board

FILAMENT TRANSFORMER

The filament transformer is a 120 volt A.C. transformer with a 3.2 volt A.C. secondary. The 3.2 volts is connected to the filament of the magnetron and results in a filament current of about 11 amps. The transformer is energized as soon as the main on-off switch is turned "on".

The filament transformer is fastened to the top of the high voltage transformer with four screws on the right side of the oven, *Figure 56*.

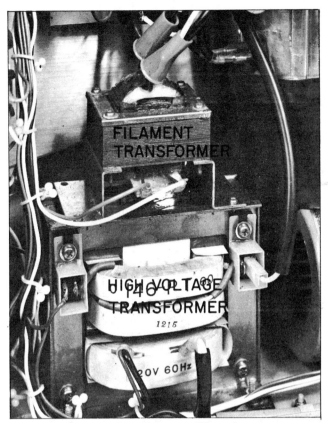

Figure 56 – Filament Transformer

The 120 volt primary winding has spade terminal connections. The secondary winding has very heavy leads and uses a *nut and bolt* connection. This connection is then covered with a slide-on insulator.

One of the secondary connections (cathode) is connected to the high voltage (3800 volts D.C.) supply.

WARNING: *NEVER TOUCH OR SERVICE THE TRANSFORMER WITHOUT UNPLUGGING OVEN AND DISCHARGING HIGH VOLTAGE CAPACITOR, TO AVOID POSSIBLE ELECTRICAL SHOCK.*

HOW TO TEST

1. Disconnect power.

2. Discharge high voltage capacitor.

3. Disconnect filament leads from magnetron or transformer.

SECTION 7 GENERAL ELECTRIC OVENS

SERVICE PROCEDURE COMPONENT DATA

4. A normal continuity check should indicate approximately 9 OHMS on the primary winding, and less than 1 OHM on the secondary winding.

5. With power "on" and primary *only* connected, the secondary should measure about 3.2 volts A.C.

HOW TO REPLACE

1. Unplug oven and discharge high voltage capacitor.

2. Disconnect leads and remove four mounting screws.

3. When reconnecting, be certain all connections are tight, and insulators are in place.

HIGH VOLTAGE TRANSFORMER

A high voltage transformer is used to provide the high voltage necessary to operate the magnetron. The transformer has a 120 volt primary winding and a secondary of about 2300 volts A.C. The 2300 volts is connected to a half-wave rectified voltage doubler circuit.

The transformer is a high leakage reactance type, which simply means that it can withstand a short circuit in the high voltage secondary circuitry without causing damage to itself or associated wiring.

The high voltage transformer is located on the right side of the oven and is bolted to the floor of the case. The filament transformer mounts on top, *Figure 56*.

HOW TO TEST

1. Unplug oven and remove outer case.

2. Discharge high voltage capacitor.

3. Remove connections from transformer terminals and check continuity. *Normal* readings should be *less than 1 OHM* on the primary, and about *70 OHMS* on the secondary.

HOW TO REPLACE

1. Unplug oven and discharge high voltage capacitor.

2. Disconnect leads and remove mounting nuts.

3. Remove filament transformer mounted on top of the high voltage transformer.

4. When reconnecting, check all leads and terminals for a tight connection.

HIGH VOLTAGE CAPACITOR

A .72 MFD 2400 volt A.C. capacitor is used along with the diodes in the voltage doubler circuit. The capacitor is mounted, by two screw to the top of the oven at the rear, *Figure 57*.

NOTE: *If the capacitor fails open, no high voltage will be available for the magnetron. A shorted capacitor should cause the house circuit fuse to blow.*

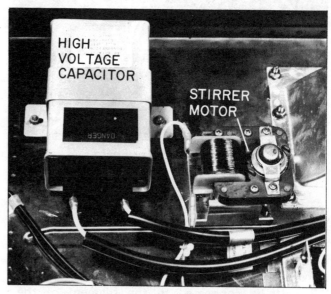

Figure 57 – High Voltage Capacitor

HOW TO TEST

1. Unplug oven and remove outer case.

2. Discharge capacitor.

SECTION 7 GENERAL ELECTRIC OVENS

SERVICE PROCEDURE COMPONENT DATA

3. Check continuity of capacitor with meter on *highest OHM scale.*

4. A *normal* capacitor will show continuity for a short time and then indicate open once the capacitor is charged.

5. A *shorted* capacitor will show continuous continuity; an *open* capacitor will show open or infinite resistance.

When replacing a high voltage capacitor, be certain to re-use the insulating spacers beteen the capacitor and the oven.

DIODES

A package of eight diodes is used in the *half-wave voltage double circuit* to rectify the high voltage to D.C. The diode board is located on the right side of the oven in front of the magnetron, *Figure 58.*

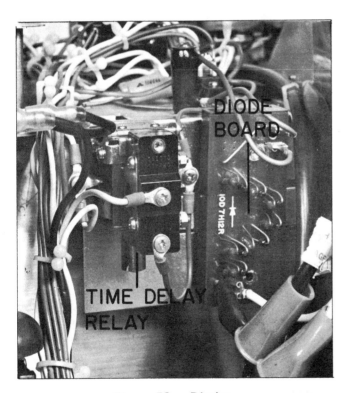

Figure 58 — Diodes

The diodes are mounted to a printed circuit board along with a one Meg OHM resistor connected across each diode. The resistors serve as current limiters during the initial surge.

Electrically, the diodes are hooked in series and serve the same function as though they were *only one large diode.* The *positive* connection is made at the top of the board, and the *negative* connection at the bottom. All connections are quick-connect or push-on terminals.

The diode board is enclosed in a phynolic case. The assembly is mounted by two *plastic screws.* If the screw heads strip, they can be removed with a pair of pliers.

HOW TO TEST

1. Unplug oven and remove outer case.

2. Discharge high voltage capacitor.

3. Isolate diodes from circuit by disconnecting leads from board.

4. *Using the highest OHM scale on meter,* check resistance between (+) and (-) terminals on diode board. Then reverse meter leads and check in both directions.

   A *normal reading* will indicate a *high* reading one direction, and a *lower* reading in the opposite direction.

5. If a short is indicated in both directions, or if an infinite resistance is read in both directions, the diode should be replaced. (Be certain to use highest OHM scale.)

HOW TO REPLACE

1. Unplug oven and remove outer case.

2. Discharge high voltage capacitor.

3. Carefully remove the plastic screws from diode board. If screws bind or strip, use a pair of pliers to loosen & remove.

CAUTION: *USE ONLY THE ORIGINAL PLASTIC TYPE SCREWS TO FASTEN THE DIODE BOARD. THE USE OF METAL SCREWS IS IMPROPER AND CAN CAUSE A SHORT CIRCUIT TO GROUND.*

DOOR LATCH SWITCH

A dual latch switch, operated by a mechanical finger level, is located at the top center of the oven. The lever is mounted inside the door and operates the latch switch as the lever is moved.

The switch assembly is actually two separate miniature pushbutton switches mounted together with screws, *Figure 59*. Each switch is actuated by a metal lever or actuator operated by the motion of the mechanical lever in the door. The lever engages a strike which prevents opening the door *without* operating the lever.

Figure 59 — Door Latch Switch

When the door is in the closed position, the lever raises the actuator and presses the switch button. This *closes* the switch contacts. When the lever is raised, the actuator releases the button and opens the contacts.

The two switches operate simultaneously but control two different circuits. One switch controls the circuit to the timer motor, stirrer motor, and cook relay coil. The other switch controls the circuit to the cook switch and the latching contacts of the cook relay.

TESTING SWITCH

1. Unplug oven and remove outer case.

2. Discharge high voltage capacitor.

3. Raise finger level on door. An audible "click" should be heard as the lever operates the switch actuator. If necessary, loosen the switch mounting screws to move the switch to insure contact with the actuator.

4. Make continuity check of each switch with door open and closed. Normal reading should indicate closed contacts when door is closed. Contacts should open as soon as finger lever is raised.

5. Check engagement of the lever with the strike in switch assembly. The strike should be adjusted for minimum play between strike and trigger when door is closed and latched. Adjust strike, if necessary, by loosening mounting screws and moving strike in correct direction, *Figure 59*.

REPLACING SWITCH

1. Unplug oven and remove outer case.

2. Discharge high voltage capacitor.

3. Open door and remove two nuts from switch bracket on back of front frame, *Figure 59*.

4. Transfer leads to new switch assembly and install.

5. Check operation according to above tests.

DOOR SAFETY SWITCH

The door safety switch acts as a backup device for the latch switch. It is an actuated pushbutton switch, rated at 125 volts, 15 amps. The switch is located on the left side of the oven, facing the door. The door hinge lever operates the switch, *Figure 60*.

The door switch is in series with the high voltage transformer primary winding. Opening the door will break the circuit to the transformer. The switch should activate within the first 1/4 of an inch door opening at the top of the door. This is a secondary safety protector if the latch switch fails.

SECTION 7 GENERAL ELECTRIC OVENS

SERVICE PROCEDURE COMPONENT DATA

Figure 60 — Door Safety Switch

Do not try to defeat the switch with a by-pass. The switch should be tested with the power off and disconnected.

TESTING THE DOOR SAFETY SWITCH

1. Disconnect the power.
2. Check continuity with the wires removed from the switch. Contacts should close when the door is closed. Contacts should open when the door is open 1/4 of an inch, as measured from the front frame and the top of the oven door, *Figure 61*.

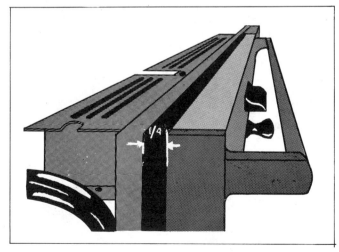

Figure 61 — Door Adjustment

3. When necessary, loosen the switch mounting screws and adjust within the safety limit specified. Tighten screws solidly.

STIRRER

A uniform heating pattern is made possible by a motor-driven stirrer blade at the top of the oven cavity, *Figure 62*. During normal operation, this area is covered with a plastic shield or cover.

Figure 62 — The Stirrer

The fan reflects the microwaves in all directions as they enter the oven cavity through the wave-guide.

The stirrer blade is driven by a motor mounted in the top of the oven. When loaded with the blade, the blade rotates at about 180 RPM.

The stirrer motor is energized through the cook relay contacts which means it is "on" *only* during cooking.

HOW TO REPLACE

1. Unplug oven and discharge high voltage capacitor.

2. Remove stirrer plastic cover from inside oven. It is fastened by three screws along the front edge.

SECTION 7 GENERAL ELECTRIC OVENS

SERVICE PROCEDURE COMPONENT DATA

3. Loosen the two screws in the blade shaft and lift out the blade.

4. Motor is mounted by three screws to top of oven. *Figure 63*.

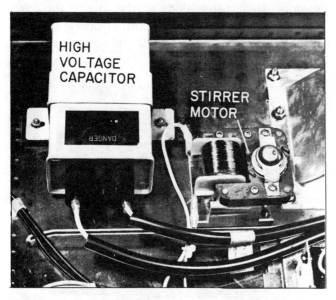

Figure 63 — Stirrer Motor

Figure 64 — Blower Motor

blown across the magnetron by means of a rubber boot or duct, *Figure 65*. Most of the air is exhausted through a vent at the rear of the case.

5. Separate the motor from nylon shaft by driving out the roll pin.

NOTE: *When reassembling, install blade with hub down and adjust so hub lines up with the end of nylon shaft, Figure.*

CAUTION: *BE CERTAIN PLASTIC STIRRER COVER IS INSTALLED PROPERLY SO STIRRER BLADE DOES NOT STALL MOTOR. FAILURE OF MOTOR AND UNEVEN COOKING CAN RESULT.*

BLOWER MOTOR

A blower motor is used to cook the magnetron and other electrical components. The blower also aids in exhausting steam and vapors from the oven cavity.

The air intake is through a filter underneath the oven inside the case, *Figure 64*. The air is then

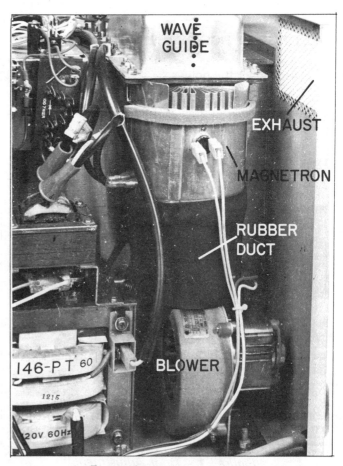

Figure 65 — Rubber Duct

SECTION 7 GENERAL ELECTRIC OVENS

SERVICE PROCEDURE COMPONENT DATA

Some air does blow into the oven cavity by means of small holes in the *wave-guide* and in the right-hand *side of the cavity* next to the magnetron. The air which passes through these openings is directed into the *stirrer cover* area and into the cavity around the rear corners of the stirrer cover. All oven vapor and steam is then exhausted through the steam ducts in the top of the outer case near the front. The steam ducts are sealed from the outer case by means of rubber gaskets glued to the inside of the outer case.

TESTING THE BLOWER MOTOR

Blower should turn "on" as soon as main on-off switch is set. Check voltage at blower connections. If the blower does not work, replace the blower assembly.

REPLACING THE BLOWER MOTOR

1. Disconnect power, remove outer case.

2. Discharge high voltage capacitor.

3. Remove rubber duct from blower.

4. Remove one mounting nut and lift blower assembly out of chassis. Disconnect leads.

5. When installing, be certain inside edge of blower bracket is under metal tab.

CAUTION: *NOTICE THAT THE BLOWER CUT-OUT IN THE BOTTOM OF THE RUBBER DUCT IS OFF-CENTER FOR PROPER AIRFLOW. INSTALL THE DUCT WITH THE CUT-OUT TOWARDS THE OVEN.*

MAGNETRON

The magnetron as shown in *Figure 60* has the following specifications:

Frequency (f)	2450 MHZ
Plate (Anode) Voltage	4.0 KV
Plate Current	280 MA
Filament Voltage	3.2 VAC
Filament Current	13A
Warm-up Time	5 SEC
Nominal Power	570 WATTS

The magnetron assembly is located on the right side of the case and is air-cooled by a blower mounted directly under the magnetron. A rubber duct connects the blower to the bottom of the magnetron assembly. A metal exhaust duct directs the air out the rear, *Figure 67.*

A wave-guide attaches to the top of the magnetron and directs the energy from the output of the tube to the oven. The wave-guide mounts to the top of the oven cavity at the rear by several nuts and bolts, *Figure 66.* A wave-guide flange on the inside of the oven provides the proper size opening, *Figure 62.* The magnetron mounts to the wave-guide by four bolts.

Figure 66 — Magnetron

The filament connects are spade type push-on terminals at the bottom of the magnetron.

WARNING: *HIGH VOLTAGE PRESENT AT FILAMENT LEADS. NEVER TOUCH OR SERVICE MAGNETRON EXCEPT AS SPECIFIED UNDER TEST INSTRUCTIONS.*

SECTION 7 GENERAL ELECTRIC OVENS

SERVICE PROCEDURE COMPONENT DATA

Figure 67 — Metal Exhaust Duct

HOW TO TEST

FOR COMPLETE MAGNETRON DIAGNOSIS, REFER TO "PLATE CURRENT TEST" AND "PERFORMANCE TEST". Continuity checks can *only* indicate an *open filament* or *shorted magnetron-type failures*. To diagnose for an *open filament* or *shorted magnetron*:

1. Disconnect power and remove outer case.

2. Discharge high voltage capacitor.

3. Isolate magnetron by disconnecting filament leads.

4. A *normal* continuity check across magnetron filament terminals should indicate less than one OHM.

5. A *normal* continuity check between terminals and chassis ground should indicate an *infinite resistance*. Little or no resistance indicates a grounded tube.

HOW TO REPLACE

1. Unplug oven and remove outer case.

2. Discharge high voltage capacitor.

3. Remove rubber blower duct and exhaust duct.

4. Remove thermal protector and filament leads.

5. While holding the magnetron, loosen and remove the four mounting nuts from the wave-guide, *Figure 68*.

Figure 68 — Removing Magnetron

6. Lower magnetron assembly slowly, until tube is clear of the wave-guide, *Figure 69*.

7. Transfer R.F. gasket to new tube, *Figure 70*.

CAUTION: *When replacing the magnetron, the R.F. gasket must be positioned correctly. Be positive that the mounting screws are tightened firmly to the wave-guide. Failure to do so can result in hazardous levels of microwave leakage.*

REMOVING THE WAVE GUIDE

1. Disconnect power to oven and remove outer case.

SECTION 7 GENERAL ELECTRIC OVENS

SERVICE PROCEDURE COMPONENT DATA

2. Discharge the high voltage capacitor.

3. Remove magnetron.

4. Remove plastic stirrer cover.

5. Remove all screws and nuts securing wave-guide and flange to top of oven cavity, *Figure*

6. When installing wave-guide, be certain to place flange inside oven as shown in *Figure 71.* Then tighten all screws and nuts.

7. Install magnetron and tighten mounting nuts.

MAGNETRON THERMAL PROTECTOR

A thermal protector is fastened to the side of the magnetron and protects the tube from overheating, such as, in the case of a blower failure, *Figure 66.*

The thermal protector is a normally closed device which opens at approximately 220°F. It resets automatically at approximately 175°F.

Figure 70 — *Installing R.F. Gasket*

Figure 71 — *Installing Wave Guide*

"Cycling of the oven lights", with no cooking, would be an indication of the thermal protector opening and closing. When this occurs, check for blower operation and obstructions in the filter under the oven.

PERFORMANCE TEST

Power output of the magnetron can be measured by performing a water temperature rise test. The standard test load for this test is two (2) quarts of water (65° - 70°). Use dish WB 64 X 68.

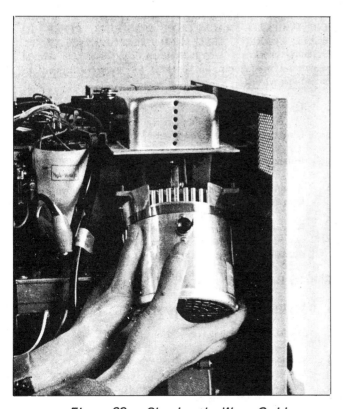

Figure 69 — *Clearing the Wave Guide*

SECTION 7 GENERAL ELECTRIC OVENS

SERVICE PROCEDURE COMPONENT DATA

The water load should be placed in the center of the glass tray and the oven turned "on" for a two (2) minute interval — (USE A STOPWATCH OR SWEEP SECOND HAND OF A WATCH — THE TIME MUST BE ACCURATE).

1. Measure line voltage (loaded) at wall outlet. (Nominal voltage is 120 volts.) Low voltage will affect temperature rise.

2. Place the WB64x68 (No. 026) round dish containing exactly two quarts of 65° - 70° water in the center of the glass tray. Record the starting water temperature exactly with an accurate glass thermometer.

3. Turn oven "on" and time cycle for exactly two minutes (120 seconds).

4. At the end of the two minute period, record the water temperature again. The difference between the starting and ending temperature is the temperature rise. (NORMAL rise is 12°F. to 16°F.)
   — If temperature rise is 12°F. - 16°F., *oven is operating normally.*
   — If no temperature rise is noted, *refer to Plate Current Test.*
   — If temperature rise is appreciably less than 12°, *check line voltage, Plate Current Test,* and *try longer cooking times.* Replace magnetron if above test fails.

WORKING VOLTAGE

As stated under specifications, the oven is designed and rated for a nominal 120 volt, 60 hertz system.

The oven will operate reliably between 105 volts and 130 volts. At the higher and lower voltages, the power output can be affected by plus or minus 15 percent

PLATE CURRENT TEST

When the *cook light* turns "on", but *little* or no heat is produced, the trouble is either in the *high voltage circuit,* or the *magnetron circuit.* In order to diagnose the problem, one or two PLATE CURRENT TESTS HAVE TO BE PERFORMED. The result of each test will indicate the source of the problem.

FIRST PLATE CURRENT TEST

1. Disconnect power and remove outer case.

2. Discharge high voltage capacitor.

3. Disconnect the grounded *test resistor green lead (A)* from the positive (+) terminal of diode board. Connect volt meter across test resistor by connecting test leads between the disconnected lead and the positive (+) terminal on diode board (observe polarity). Set volt meter to 10-12 volt D.C. scale, *Figure 72.*

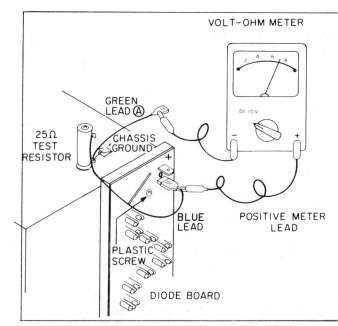

Figure 72 — Testing First Plate Current

4. Plug in oven, use water load, and set oven for cooking.

5. Observe volt meter reading and follow *FIRST PLATE CURRENT DIAGNOSIS CHART* below.

SECOND PLATE CURRENT TEST

(Use this test only if indicated by FIRST PLATE CURRENT TEST)

1. Disconnect power.

2. Discharge high voltage capacitor.

SECTION 7 GENERAL ELECTRIC OVENS

SERVICE PROCEDURE COMPONENT DATA

3. Using a jumper wire with alligator clips, connect one end to negative (-) terminal on diode board. Connect other end to chassis ground, *Figure 73*.

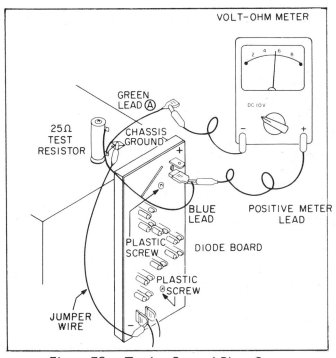

Figure 73 — Testing Second Plate Current

4. Disconnect the grounded *test resistor green lead (A)* from the positive (3) terminal of diode board. Connect volt meter between the disconnected lead and the positive (+) terminal on diode board (observe polarity). Set to 10-12 volt D.C. scale, *Figure 73*.

5. Plug in oven, use water load, and set oven for cooking.

6. Observe volt meter reading and follow *SECOND PLATE CURRENT DIAGNOSIS CHART*

FIRST PLATE CURRENT DIAGNOSIS CHART

D.C. Test Voltage	Diagnosis and Procedure
7.0 Volts (Approx.)	Test voltage 7.0 volts indicates *normal* plate current (280 MA) RUN PERFORMANCE TEST.
5.5 Volts (Approx.)	1. Short Circuit to ground someplace between voltage doubler circuit and magnetron. *CHECK WIRING.* 2. Shorted magnetron tube *TEST TUBE FOR SHORT.*
0 Volts	PERFORM SECOND PLATE CURRENT TEST

SECOND PLATE CURRENT DIAGNOSIS CHART

D.C. Test Voltage	Diagnosis and Procedure
5.5 Volts	1. Magnetron tube filament open — *TEST TUBE FOR OPEN CIRCUIT.* 2. Filament transformer open — *TEST TRANSFORMER.* 3. Open wiring to magnetron or filament transformer.
0 Volts	1. High voltage transformer open or shorted — *TEST TRANSFORMER.* 2. Diode board open or shorted — *TEST DIODE BOARD.* 3. High voltage capacitor open or shorted — *TEST CAPACITOR.*

SECTION 7 GENERAL ELECTRIC OVENS

SERVICE PROCEDURE COMPONENT DATA

DOOR

The door is bottom-hinged and contains a screen window with a tempered glass inside and a full size plastic window outside. Each of two hinge levers employ a door spring which hooks into a hole in the lever and to the rear of the case, *Figure 74*. The main body of the door is a die-cast picture frame to which the outer window, or door trim, handle, and door seals are fastened. To this main door assembly is mounted an inner door assembly called the CHOKE SEAL. This is the door "plug" which extends inside the oven cavity and consists of the inner window, inner door, poly seal, and seal plate.

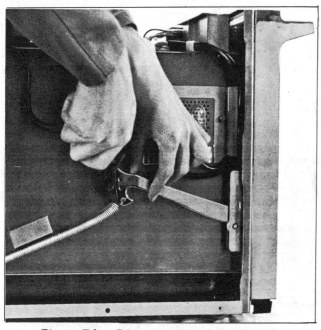

Figure 74 — Disconnecting Door Springs

DOOR SEALS

The primary door seal is accomplished by use of a CHOKE SEAL. With this type of seal, there is no metal-to-metal contact between the door and the oven. Instead, a scientifically designed gap between the "plug" and the oven provides an electronic barrier to microwaves. This gap and the construction of the door in this area effectively creates a short circuit path for microwaves and thereby prevents their escape.

To improve the seal, a secondary absorber seal is used to absorb any leakage which might get past the CHOKE SEAL. The absorber seal is inside the main part of the door which seals against the front frame just outside the oven cavity. It consists of strips of ferrite impregnated rubber at the top, sides, and bottom of the door. A gray plastic cover is placed over the ferrite for wear and appearance, *Figure 75*.

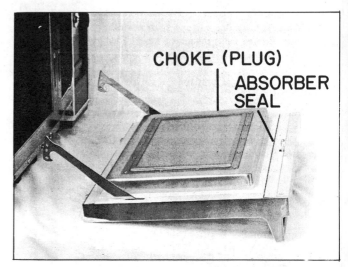

Figure 75 — Choke and Absorber Seal

DOOR ADJUSTMENT

DOOR ADJUSTMENT

Door adjustment is limited to a slight side-to-side adjustment of the hinge which fastens to the front frame. When checking the door, make certain hinge mounting screws are tight. Open and close door while observing gap between door plug and oven cavity as door approaches full closure. This gap should be about 1/8". The corners of the door can be checked by insertion of a narrow paper strip over the corner of the inner door "plug" and pulling out the paper while holding the door open slightly. The paper should move freely, which indicates no mechanical interference.

If necessary, the door can be moved side-to-side slightly by loosening the hinge screws on the front frame, *Figure 76*. Move the door and retighten the screws.

SECTION 7 GENERAL ELECTRIC OVENS

SERVICE PROCEDURE COMPONENT DATA

TO REMOVE THE DOOR

1. Unplug oven and remove outer case.

2. Discharge high voltage capacitor.

3. Unhook door spring on each side.

4. Remove the bottom door trim (three screws).

5. Remove left-hand oven hinge, *Figure 76* by taking out the two oval-head screws.

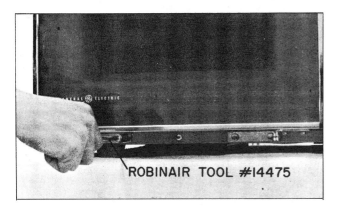

Figure 76 — Removing Left Side Hinge

6. While holding door, trigger latch and carefully slide door to right to free from remaining hinge. Then guide hinge levers out of opening in front frame, *Figure 77*.

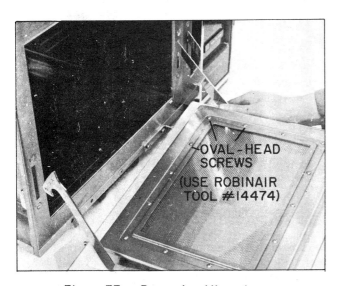

Figure 77 — Removing Hinge Lever Roller Assembly

7. The nylon hinge lever roller assembly can then be removed, *Figure 74*.

NOTE: *After door has been reassembled to oven, be certain to check door adjustment and interlock system as outlined.*

TO DISASSEMBLE DOOR

1. Remove door and lay on flat surface.

2. Remove all oval-head screws, *Figure 77*, around inside window. (Use special tool Robinair No. 14474.)

3. Lift off the inner door assembly (*Figure 78*). This consists of the white seal plate, the poly seal, door back plate (plug), and window. The assembly is fastened together by 28 screws.

Figure 78 — Removing Door Window

4. The ferrite sealer assemblies at the top, bottom and sides of the door can be removed by taking out three screws in each. The side seals must be removed first.

NOTE: *Each sealer consists of a base plate, ferrite rubber seal, plastic seal cover, and holder or retainer strip, Figure 79.*

5. Remove the hinge levers mounted to door casting by two screws.

117

SECTION 7 GENERAL ELECTRIC OVENS

SERVICE PROCEDURE COMPONENT DATA

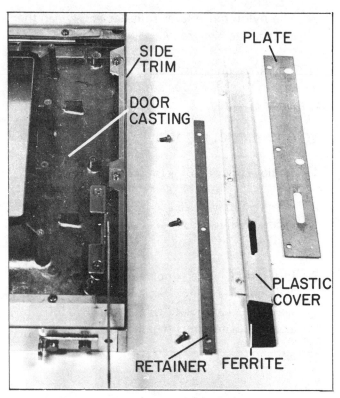

Figure 79 — Sealer Components

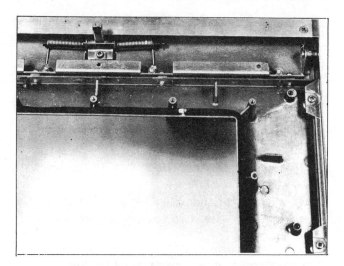

Figure 80 — Removing Catch Lever

6. Remove side die-cast trim from each side of door by removing five screws from each side and one small screw from underside of handle at each end, *Figure 80.*

   NOTE: *Three screws fasten in the door casting and one each to the top trim and latch head mounting plate. Then pull side trim loose.*

7. Bottom trim can then be pulled off.

8. Remove latch lever by taking out one screw mounting lever to latch head. Hold down latch head trigger and work lever out of opening, *Figure 80.*

9. Remove top die-cast trim. Trim is attached to latch bracket by four fasteners and nuts. Loosen the nuts and slide the trim to one side until free from the fasteners, *Figure 81.*

10. The latch spring can be removed by pulling off the keeper at the end of the shaft. Then slide the shaft out of the spring.

    NOTE: *When installing new spring, be certain to mount as shown in, Figure 80.*

Figure 81 — Removing Die Cast Trim

11. To reassemble the top trim, slide the trim over the fasteners one at a time, while guiding each fastener into the groove on the trim. Center the trim and tighten the fastener nuts.

12. Reassemble remainder of door.

13. Install hinge levers with "hook" down.

    NOTE: *After door has been reassembled to oven, be certain to check door seal and interlock system as outlined.*

SECTION 8 LITTON MICROWAVE OVENS

SERVICE PROCEDURE COMPONENT DATA

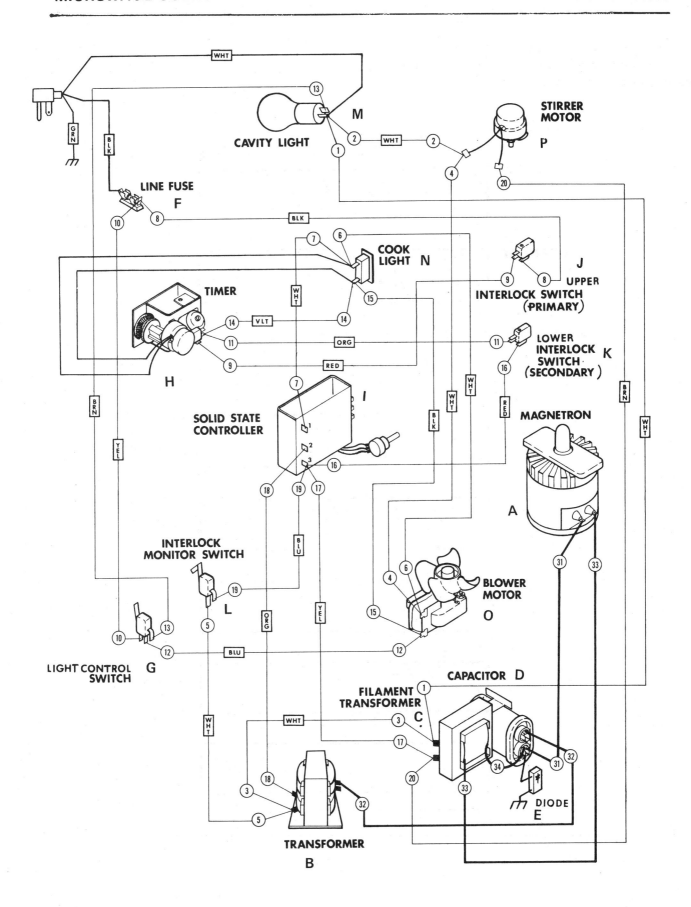

Figure 82 - A thru P

SECTION 8 LITTON MICROWAVE OVENS

SERVICE PROCEDURE COMPONENT DATA

The following chart shows the per cent of time power is supplied to the power transformer during the various cooking modes, *Figure 83*.

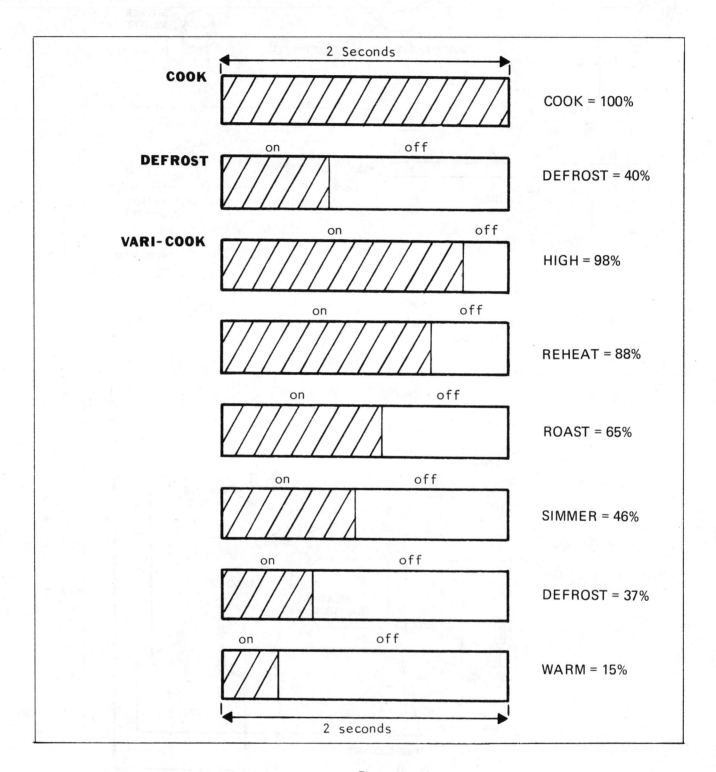

Figure 83

SECTION 8 LITTON MICROWAVE OVENS

SERVICE PROCEDURE COMPONENT DATA

WARNING TO SERVICE TECHNICIANS

The Model 416.000 microwave oven has a monitoring system designed to assure proper operation of the safety interlock system.

The interlock monitor switch will immediately cause the oven fuse to blow if the door is opened while the following *combined* failures exist:

1. Upper and lower door interlock switch contacts failed in a closed position.
2. Timer contacts in a closed position.

CAUTION: Replace blown fuse with 15 ampere fuse only.

Before replacing the blown oven fuse, test the upper and lower door interlock switches and interlock monitor switch for proper operation as described in the component test procedures.

Do not attempt to repair sticking contacts of any interlock switch.

Any indication of sticking contacts during component tests requires replacement of that component to assure reliability of safety interlock system.

CAUTION

To avoid possible exposure to microwave radiation or energy, visually check the oven for damage to the door and door seal before operating any oven. Use your microwave survey meter to check the amount of leakage before servicing. In the event that the R.F. leakage exceeds 5 mw/cm^2 at 5 cm., appropriate repair must be made before continuing to service the unit. Check interlock function by operating the door latch. The oven cook cycle should cut off before the door can be opened.

The door and latching assembly contain the radio frequency energy within the oven. The door is protected by three safety interlock switches. Do not attempt to defeat them. Under no circumstances should you try to operate the oven with the door open.

Proper operation of the microwave ovens requires that the magnetron be properly assembled to the waveguide and cavity. Never operate the magnetron unless it is properly installed.

Untrained personnel should not attempt service without a thorough review of the test procedures and safety information contained in this manual.

SECTION 8 LITTON MICROWAVE OVENS

SERVICE PROCEDURE COMPONENT DATA

TABLE OF CONTENTS

SECTION A General Information

SECTION B Oven Specifications and Electrical Component Features

SECTION C Function of Electrical Components

SECTION D Operating Sequence

SECTION E Troubleshooting Chart

SECTION F Component Test Procedures

SECTION G Disassembly and Replacement of Components

SECTION H Wiring Schematic

SECTION 8 LITTON MICROWAVE OVENS

SERVICE PROCEDURE COMPONENT DATA

SECTION A: GENERAL INFORMATION

MICROWAVE COOKING

Litton microwave ovens use microwave energy to heat or cook food in a fraction of the time needed to cook with conventional ovens. Unlike conventional ovens, a microwave oven heats food without applying external heat.

A magnetron tube is used, *Figure 82-A,* to produce short, electromagnetic waves known as microwaves, or R.F. energy. This energy is directed into the cooking cavity where the food is placed to be heated.

The microwaves readily pass through many materials, such as glass, most plastics, paper and china, with little or no effect. Generally, these materials make excellent utensils for cooking with microwaves.

Some other materials, such as metal and foil, tend to reflect microwave energy. Except for certain recommended procedures for use of metal or foil as outlined in Litton cookbooks or instruction manuals, use of metal utensils or utensils with metal parts should be avoided in microwave ovens for the following reasons:

1. Metal utensils do not allow complete penetration of the food by the microwaves, so oven efficiency is greatly reduced. Foods heated in open metal utensils heat much slower than if heated in materials suited for microwave heating.

2. If the cooking load is not large enough to absorb the microwave energy, the oven could be damaged by an arc between the metal utensils and the cavity interior or door assembly.

3. Magnetron tube life can be shortened by extended periods of back-feeding R.F. energy which raises the magnetron filament temperature.

Because metal reflects microwave energy, the metal walls of the cooking cavity are not affected by microwaves and do not get hot.

Materials with high moisture content, like most foods, absorb microwave energy. As the electromagnetic waves at a frequency of 2450 megacycles enter the food, the molecules tend to align themselves with the energy. Since the microwaves are changing polarity with every half cycle, the food molecules follow suit and thus move rapidly back and forth. Effectively, the food molecules are changing direction every half cycle, so they are oscillating back and forth 4,900,000,000 times each second. This high speed oscillation causes friction between the molecules, thereby converting the microwave energy to heat.

SECTION B: OVEN SPECIFICATIONS AND ELECTRICAL COMPONENT FEATURES

OVEN SPECIFICATIONS

Power Source

110 to 125 volts a-c
15 amperes (single circuit)
60 cycles per second
Single phase, 3 wire grounded

Power Output

Cook: 600 watts nominal of r. f. microwave energy Operating frequency of 2450 MHz

Vari-cook: 80 – 600 watts nominal of r. f. microwave energy.

Power Consumption

OFF Condition – None
Cavity Light On Condition – 40 watts
COOK Condition – 1500 Watts

ELECTRICAL COMPONENT FEATURES

HIGH VOLTAGE COMPONENTS (*Half Wave Doubler System*)

1. Magnetron, *Figure 82-A.*
2. Power Transformer, *Figure 82-B.*
3. Filament Transformer, *Figure 82-C.*
4. Capacitor, *Figure 82-D.*
5. Diode, *Figure 82-E.*

SECTION 8 LITTON MICROWAVE OVENS

SERVICE PROCEDURE COMPONENT DATA

CONTROL CIRCUITRY COMPONENTS

1. Fuse, *Figure 82-F.*
2. Light Control Switch, *Figure 82-G.*
3. Digital Timer Assembly (59 min., 55 sec.), *Figure 82-H.*
4. Solid State Controller, *Figure 82-I.*

DOOR SAFETY INTERLOCK SWITCHES

1. Upper Door Interlock Switch (primary), *Figure 82-J.*
2. Lower Door Interlock Switch (secondary), *Figure 82-K.*
3. Interlock Monitor Switch, *Figure 82-L.*

LIGHTS

1. Cavity Light, *Figure 82-M.*
2. Cook Light, *Figure 82-N.*

MOTORS

1. Blower Motor, *Figure 82-O.*
2. Stirrer Motor, *Figure 82-P.*

SECTION C: FUNCTION OF ELECTRICAL COMPONENTS, *Figure 82*

FUSE (F)

The oven is protected from interlock switch failures and current overloads by a 15 ampere ceramic fuse.

MAGNETRON TUBE (A)

The microwave energy used to heat or cook foods in the oven is produced by the magnetron tube.

The basic magnetron tube is a vacuum tube with a cylindrical cathode within a cylindrical anode surrounded by a magnetic field. When the magnetron filaments are heated by the transformer filament secondary voltage, and the high negative d-c voltage is applied to the tube cathode, negatively charged electrons are emitted from the cathode and attracted to the more positive anode of the tube.

Ordinarily, the electrons would travel in a straight line from the cathode to the anode. However, the magnetic field, provided by the permanent magnets surrounding the anode, causes the electrons to take an orbital path between the cathode and the anode, see text Frigidaire microwave oven, *Figure C.*

As the electrons approach the anode, they travel past the small resonant cavities that are part of the anode. Interactions occur, causing the resonant cavities to oscillate at a very high frequency of 2450 megacycles.

This r. f. energy is radiated from the magnetron antenna into the waveguide, into the cooking cavity feedbox, past the stirrer blade assembly and finally into the cooking cavity where food is placed to be heated.

POWER TRANSFORMER (B)

The purpose of the power transformer is to provide the high voltage needed for magnetron tube oscillation.

During a cook cycle, the 120 volts a-c applied to the primary winding is converted to approximately 4000 volts peak to peak a-c on the high voltage secondary winding of the power transformer during normal operation.

FILAMENT TRANSFORMER (C)

The purpose of the filament transformer is to provide voltage for heating the magnetron filaments.

When the timer is set to a selected time setting and the door is closed, the 120 volts a-c applied to the primary winding of the filament transformer is converted to approximately 3 volts a-c on the secondary winding.

CAPACITOR (D)

The capacitor is part of the half-wave voltage doubler system that converts the high a-c voltage produced by the power transformer to the approximate 4000 volts negative d-c necessary for magnetron operation.

SECTION 8 LITTON MICROWAVE OVENS

SERVICE PROCEDURE COMPONENT DATA

The capacitor stores electrical energy during alternate half cycles of high a-c voltage produced by the power transformer.

DIODE (E)

The diode is a solid state device that allows current flow in one direction, but prevents current flow in the opposite direction. When the diode is used in conjunction with the capacitor, a voltage doubling circuit is provided that converts the high a-c voltage from the power transformer to the approximate 4000 volts negative d-c needed for magnetron operation.

LIGHT CONTROL SWITCH (G)

With the door open, the light switch completes a circuit to the cavity light even with the timer in the OFF position.

With the door closed, the light switch provides a circuit to the cavity light only if a cooking time is selected on the timer. The light switch is activated by the door hook alignment pin pressing against the light control switch arm.

BLOWER MOTOR (O)

The blower motor drives a blower fan which draws cooling air through the oven base. This cooking air directed through the air ducts surrounding the magnetron tube to cool the magnetron. Most of the air is then exhausted directly through the back vents.

However, a portion of this air is channeled through the cavity to remove steam and vapors given off from the heating food. It is then directed through an air duct and exhausted from vents in the top of the outer case.

STIRRER MOTOR (P)

The stirrer motor turns the stirrer blades at the top of the oven cavity. The stirrer blades revolve at 60 r.p.m. and reflect the electromagnetic energy produced by the magnetron tube. This reflected energy bounces back and forth between the walls, top and floor of the metal cooking cavity. This allows the r.f. energy to penetrate the food from all sides to give a uniform heating pattern.

DOOR INTERLOCK SWITCHES (Upper and Lower) (J and K)

The upper and lower door interlock switches are mounted on a latch slide assembly that moves up when the door is closed. The switches are activated by the door hooks.

When the door is open, the door interlock switches interrupt voltage to all components except the cavity light. Since the power transformer receives no voltage with the switches open, a cook cycle cannot occur until the door is firmly closed and both door interlock switches are activated.

DIGITAL TIMER (H)

Desired cooking times from 0 to 59 minutes, 55 seconds, may be selected by turning the timer knob. Cooking times in tens of minutes may be selected by rotating the timer thumbwheel.

Pushing the timer knob in and rotating activates the timer switch contacts, which open or close the circuit to the lower door interlock switch, and various other components.

The timer motor begins operation when the door is closed. When the timer nears the end of the selected cooking time, the timer bell rings once before the timer switch is deactivated. Since no electrical power is allowed past the open timer switch, the oven reverts to the OFF condition.

INTERLOCK MONITOR SWITCH (L)

The interlock monitor switch mounted on the oven cavity near the cavity rollers is activated by the door hook alignment pin when the door is closed.

The interlock monitor switch will immediately cause the oven fuse to blow if the door is opened while the following *combined* failures or conditions exist.

1. Upper and lower door interlock switch contacts failed in a closed position.

2. Timer contacts in a closed position.

CAUTION: *REPLACE BLOWN FUSE WITH 15 AMPERE FUSE ONLY. (F)*

SECTION 8 LITTON MICROWAVE OVENS

SERVICE PROCEDURE COMPONENT DATA

Before replacing the blown oven fuse, test the upper and lower door interlock switches, timer contacts, and interlock monitor switch for proper operation as described in the COMPONENT TEST PROCEDURES.

DO NOT ATTEMPT TO REPAIR STICKING CONTACTS OF ANY INTERLOCK SWITCH.

Any indication of sticking contacts during component tests requires replacement of that component to assure reliability of the safety interlock system.

CAVITY LIGHT (M)

The cavity light illuminates the interior of the oven cavity so that food being heated can be examined visually through the door window without having to open the door.

COOK LIGHT (N)

The cooking light located on the control panel illuminates to give a visible indication that the oven is in a cook cycle.

SOLID STATE CONTROLLER

The solid state controller varies the 60 Hz. a-c input to the power transformer within a two-second time base. For example, 10% of full power would be 0.2 seconds of 60 Hz. a-c, and 1.8 seconds of OFF time.

SECTION D: OPERATING SEQUENCE

OFF CONDITION (*Door CLOSED, Timer OFF*)

With the door closed and the timer in the OFF position, no components operate.

CAVITY LIGHT ON CONDITION (*Door OPEN*)

When the door is opened, the light control switch completes the circuit to illuminate the cavity light only.

COOK CONDITION (*Door CLOSED, Timer ON*)

To put the oven into a COOK condition, the door must be closed securely. This activates three door interlock switches and the light control switch. The upper and lower door interlock switches, mounted on the latch slide bracket, are activated by the door hooks. The interlock monitor switch and light control switch are activated by the door hook alignment pin.

When the desired cooking time is selected on the digital timer, the timer switch contacts close, and the circuit is completed to the following components:

1. The blower motor forces cooking air through the air ducts around the magnetron which forces a portion of this air through the cooking cavity.

2. The cavity light illuminates the cavity interior.

3. The stirrer motor rotates the stirrer blades at the top of the oven cavity.

4. The cook light illuminates to indicate that the oven is in a cook cycle.

5. The timer motor begins timing out the selected cooking time.

6. The 120-volt a-c input to the primary winding of the filament transformer is converted to approximately 3 volts a-c to heat the magnetron filaments.

7. The 4000 volts a-c output from the power transformer high voltage secondary winding is sent to a half-wave voltage doubler circuit comprised of a single diode and one capacitor where approximately 4000 volts negative d-c (peak voltage) is provided to the magnetron tube assembly.

8. The negative 4000 volts d-c applied to the cathode of the magnetron tube causes the magnetron to oscillate and produce the 2450 MHz cooking frequency.

9. The R.F. energy produced by the magnetron antenna is channeled through a waveguide, into the cavity feedbox, past the stirrer blade, and finally into the cavity, where the food is placed to be heated.

10. Upon completion of the selected cooking time, the timer bell rings once as the timer switch is deactivated. Since no electrical power is allowed past the open timer switch, the oven reverts to the OFF condition.

SECTION 8 LITTON MICROWAVE OVENS

SERVICE PROCEDURE COMPONENT DATA

SECTION E: MICROWAVE OVEN TROUBLE-SHOOTING CHART

PROBLEM	POSSIBLE CAUSE
Totally inoperative.	1. Oven power cord not connected. 2. Blown wall fuse or tripped breaker. 3. Blown oven fuse. 4. Broken or loose wire connection within power cord or wire harness.
Line fuse blows or circuit breaker trips when oven is put into a cook cycle.	1. Overloaded circuit. 2. Shorted component or harness wire.
Line fuse blows when oven is in defrost or vari-cook condition.	1. Solid state controller defective. 2. Shorted component or harness wire.
Oven fuse blows when door is opened.	1. Upper door interlock switch and lower door interlock switch defective. 2. Interlock monitor switch defective. 3. Shorted harness wire.
Cavity lamp does not illuminate; other oven operation normal.	1. Loose or burned out cavity lamp. 2. Defective lamp socket. 3. Defective light control switch. 4. Broken or loose wire connection.
Cavity lamp illuminates with door open, but light goes out when door is CLOSED with timer ON. No components operate and oven will not heat.	1. Upper door interlock switch defective or out of adjustment. 2. Defective timer switch. 3. Broken or loose wire connection.
Cook cycle operates normally, but with DEFROST button depressed, oven does not heat.	1. Defective solid state controller. 2. Broken or loose wire connection.
Cook cycle and defrost cycle operate normally, but with vari-cook button depressed, oven does not heat.	1. Defective solid state controller. 2. Broken or loose wire connection.
Cook cycle and defrost cycle operate normally, but when vari-cook control is rotated, cook cycle does not vary.	1. Vari-cook control defective. 2. Broken or loose wire connection.
Cook light does not illuminate but other operation normal.	1. Defective cook light. 2. Broken or loose wire connection.

SECTION 8 LITTON MICROWAVE OVENS
SERVICE PROCEDURE COMPONENT DATA

Magnetron blower motor does not operate, other oven operation normal.	1. Defective blower motor. 2. Blower fan binding. 3. Broken or loose wire connection.
Cavity light varies in intensity and oven makes a slight humming sound when oven is in defrost or vari-cook operation.	1. This is a normal indication when the oven is operating properly.
Digital timer does not advance with oven in cook cycle. Oven heats normally.	1. Defective timer motor. 2. Binding timer assembly. 3. Broken or loose wire connection.
Stirrer blades do not turn at top of oven cavity during cook cycle.	1. Defective stirrer motor. 2. Stirrer blades binding. 3. Broken or loose wire connection.

Oven appears to operate normally, but heats very slowly.
 	
NOTE: *If test for components a through c indicates normal, replace magnetron and check oven for proper operation.*
	1. Vari-cook button depressed and control turned to warm.
	2. Low line voltage (should be at least 110 volts a-c).
	3. Defective solid state controller.
	4. Problem in high voltage section. See specific tests for high voltage components.
	
	   a. Magnetron
	   b. Capacitor
	   c. Power Transformer

Oven appears to operate normally, but does not heat at all.
 	
NOTE: *If test for components a through e indicates normal, replace magnetron and check oven for proper operation.*
	1. Lower door interlock switch defective or out of adjustment.
	2. Defective solid state controller.
	3. Problem in high voltage section. See specific tests for high voltage components.
	
	   a. Diode
	   b. Magnetron
	   c. Capacitor
	   d. Power Transformer
	   e. Filament Transformer
	
	4. Broken or loose wire connection.

SECTION 8 LITTON MICROWAVE OVENS

SERVICE PROCEDURE COMPONENT DATA

SECTION F: COMPONENT TEST PROCEDURES

HIGH VOLTAGE COMPONENT TEST

The following procedures should be followed when little or no heat is produced by the oven, but all other operations including cook light operation appear normal. This operation should be tested with the COOK button depressed.

CAUTION: *HIGH VOLTAGES ARE PRESENT DURING THE COOK CYCLE. EXTREME CAUTION SHOULD BE OBSERVED AT ALL TIMES.*

IT IS NEITHER NECESSARY NOR ADVISABLE TO ATTEMPT MEASUREMENT OF HIGH VOLTAGES.

DO NOT TOUCH ANY OVEN COMPONENTS, OR WIRING, DURING OVEN OPERATION. ATTACH METER LEADS WITH ALLIGATOR CLIPS WHEN MAKING OPERATIONAL TESTS.

BEFORE TOUCHING ANY OVEN COMPONENTS OR WIRING, ALWAYS UNPLUG THE OVEN FROM ITS POWER SOURCE AND DISCHARGE THE CAPACITOR BY SHORTING ACROSS THE CAPACITOR TERMINALS WITH AN INSULATED SCREWDRIVER.

DIODE TEST (E)

The diode is located on the back of the base near the capacitor. Since the diode is installed as a parallel circuit to the magnetron tube, the diode circuit and the tube circuit must be isolated to determine where the problem exists.

1. Unplug oven power cord, remove the outer case, and discharge the capacitor.

2. Visually check the high-voltage wiring for proper connections.

3. Isolate the diode from the tube circuit by disconnecting the diode from the capacitor.

4. *With the ohmmeter set on the highest resistance scale, measure the resistance across the two diode terminals. Reverse the meter leads and again observe the resistance reading.

A normal diode should read infinite resistance or open in one direction, and approximately 50,000 ohms or more when the meter leads are reversed.

Procedure (a): If the diode check is normal, re-connect the diode and go to Magnetron Test in this section.

Procedure (b): If continuity is indicated in both directions, or if an infinite resistance is read in both directions, the diode is probably defective and should be replaced.

*NOTE: *Meters with less than a 6-volt battery are usually not adequate for checking the front-to-back resistance of the dioe. The meter should be checked with a diode known to tbe good before judging a diode to be defective since low battery voltage can give an indication of infinite resistance in both directions.*

When installing the diode, it is important to observe the correct polarity (indicated by arrow or red dot) of the diode. The arrow or red dot should point toward the oven base.

The opposite end of the diode should be connected to the capacitor. When mounting the diode, keep the diode lead and body well away from metal parts to prevent possible arcing.

MAGNETRON TEST (A)

1. Unplug the power cord, remove the outer case and discharge the capacitor.

2. Disconnect the high voltage leads from the magnetron filament terminals.

3. Measure the resistance across the two magnetron filament terminals with the ohmmeter set on its lowest resistance scale (R X 1).

Procedure (a): If a normal resistance of less than 1 ohm between magnetron terminals is indicated on the meter, go on to Step 4.

Procedure (b): If high resistance or infinite resistance is indicated between the magnetron terminals, replace the magnetron assembly and check the oven for proper operation.

SECTION 8 LITTON MICROWAVE OVENS

SERVICE PROCEDURE COMPONENT DATA

With the ohmmeter set on the highest resistance scale, measure the resistance between each magnetron filament terminal and magnetron chassis ground.

Procedure (a): If the ohmmeter indicates infinite resistance (regardless of meter polarity), re-connect the high voltage lead to the magnetron and go on to CAPACITOR TEST.

Procedure (b): If continuity is read with the meter, replace the magnetron assembly and check the oven for proper operation.

CAPACITOR TEST (D) *(Isolate the component before making the continuity tests.)*

If the capacitor is open, no high voltage is available to the magnetron. A shorted capacitor normally causes high line current, which should trip the wall circuit breaker or blow the line fuse.

An ohmmeter can be used to check for a shorted or open capacitor.

1. Unplug the oven, discharge the capacitor, and remove the leads from the capacitor terminals.

2. With an ohmmeter set on the highest scale, measure the resistance between the two capacitor terminals. The meter needle should momentarily deflect upward to indicate continuity and should then return to infinity. Reversing the meter leads should give the same indication.

3. If the ohmmeter indicates continuity between the capacitor terminals at all times, or if no meter deflection occurs at all, the capacitor should be replaced.

NOTE: *The ohmmeter should have a minimum 6-volt battery voltage and should be set on the high-resistance scale when making a capacitor check.*

FILAMENT TRANSFORMER TESTS (C)

An operational test can be made in order to determine the amount of filament voltage produced by the filament transformer.

CAUTION: *HIGH VOLTAGE UP TO 6100 VOLTS A-C CAN BE PRESENT AT THE HIGH VOLTAGE TERMINAL OF THE POWER TRANSFORMER DURING A COOK CYCLE. OBSERVE CAUTION AT ALL TIMES.*

1. With oven power cord unplugged and outer case removed, discharge the capacitor with an insulated screwdriver.
2. Completely remove the HIGH VOLTAGE LEAD that connects the capacitor to one of the high-voltage terminals (marked "Hi" or "Lo".)

CAUTION: *DURING TEST OPERATION, KEEP METER LEADS, HANDS, ETC., WELL AWAY FROM THE HIGH-VOLTAGE TERMINALS OF THE POWER TRANSFORMER.*

3. With alligator clips, connect a meter capable of measuring up to 10 volts a-c across the magnetron filament terminals.

4. Apply power to the oven and put the oven into a cook cycle to get a filament voltage reading. A normal indication should be approximately 3 volts a-c.

5. Unplug the oven and disconnect the meter leads.

Procedure (a): If a normal 3 volts a-c reading was indicated on the meter, go on to Power Transformer Test.

Procedure (b): If no voltage was indicated on the meter, set the meter to the proper scale and check to see if the 120-volt a-c primary input to the filament transformer is present during a cook cycle. If the input voltage is normal, but no filament secondary voltage is present, replace the filament transformer.

POWER TRANSFORMER TEST

One side of the power transformer high voltage secondary winding is connected to oven chassis ground. The other end of the high voltage secondary has two terminals marked "Hi" and "Lo" to provide a method of increasing or decreasing the amount of power produced by the magnetron.

SECTION 8 LITTON MICROWAVE OVENS

SERVICE PROCEDURE COMPONENT DATA

CAUTION: *HIGH VOLTAGE UP TO 6100 VOLTS A-C CAN BE PRESENT AT THE HIGH VOLTAGE SECONDARY TERMINALS DURING A COOK CYCLE. IT IS NOT RECOMMENDED THAT HIGH VOLTAGE MEASUREMENTS BE ATTEMPTED.*

Normally, a continuity check of the high voltage secondary winding will be sufficient to determine the condition of the secondary winding.

1. With oven unplugged, discharge the capacitor and disconnect the high-voltage lead from the power transformer high voltage secondary terminal.

2. With an ohmmeter set on the low resistance scale, (R X 1) measure the resistance between each high-voltage terminal (marked "Hi" and "Lo") of the power transformer and oven chassis ground. The meter should indicate approximately 75 ohms between oven chassis ground and the transformer high-voltage terminal marked "Lo," and should indicate approximately 80 ohms between oven chassis ground and the transformer high-voltage terminal marked "Hi."

3. If extremely low resistance is indicated, or if extremely high or infinite resistance is indicated with the continuity test, replace the power transformer.

SOLID STATE CONTROLLER TEST (I)

An operational test can be made to determine if the solid state controller is providing proper input to the power transformer.

1. Unplug the power cord, remove the outer wrap, AND DISCHARGE THE CAPACITOR.

CAUTION: *HIGH VOLTAGES UP TO 6100 VOLTS A-C CAN BE PRESENT AT THE HIGH-VOLTAGE TERMINAL OF THE POWER TRANSFORMER DURING A COOK CYCLE. OBSERVE CAUTION AT ALL TIMES.*

2. With alligator clips, connect a meter capable of measuring up to 125 volts a-c across terminal #1 and terminal #3. (Press cook button and turn timer on.)

Procedure (a): If a constant 110-125 volts a-c reading was indicated on the meter, go to Step 3.

Procedure (b): If no voltage was indicated on the meter, or the meter needle fluctuates during this est, replace the solid-state controller.

3. TURN TIMER OFF AND DISCHARGE THE CAPACITOR. With alligator clips, connect meter across terminal #1 and terminal #2. (Press vari-cook button and turn timer on.)

NOTE: *When performing the test, turn vari-cook control to each setting — high, reheat, roast, simmer, defrost, warm.*

Procedure (a): If the meter needle fluctuates during each control setting, go to Step 5.

Procedure (b): If 0 voltage is indicated or a constant 120 volts is indicated during any control setting, replace the solid-state controller.

5. If Step 2 through 4 indicates normal readings and thee oven produces little or no heat, check the high-voltage components.

DOOR INTERLOCK SWITCH TEST (*Upper/Lower*)

The upper and lower door interlock switches are mounted on a latch slide assembly which moves up when the door is closed. As the slide moves up, the switches are activated by the latch hooks mounted on the door assembly.

The door interlock switches are mounted on adjustable strikes which should be aligned to provide maximum switch activation by the door hooks when the door is closed.

The switches can be tested with an ohmmeter.

1. Unplug the oven power cord, remove outer case and discharge the capacitor.

2. Disconnect the harness leads from the switch terminals.

3. With the oven door closed, use an ohmmeter to check continuity between the switch terminals marked NO (Normally Open) and COM (Com-

SECTION 8 LITTON MICROWAVE OVENS

SERVICE PROCEDURE COMPONENT DATA

mon). Since the switches are activated with the door closed, the ohmmeter should indicate continuity.

4. Open the oven door and check continuity between terminals NO (Normally Open) and COM (Common). The ohmmeter should indicate an open circuit with the door opened.

5. If improper indications are given, make necessary switch adjustments or replace the defective switch and re-connect the harness leads to the switch terminals.

INTERLOCK MONITOR SWITCH TEST

The interlock monitor switch mounted on a bracket near the cavity rollers is designed to assure the proper operation of the safety interlock system.

The interlock monitor switch can be tested with an ohmmeter.

1. Unplug the oven power cord, remove outer case and discharge the capacitor.

2. Disconnect the harness leads from the interlock monitor switch.

3. With the door closed, a continuity check between the switch terminal marked COM (Common) and the switch terminal marked NC (Normally Closed) should indicate an open circuit.

4. Open the oven door.

A continuity check between the COM terminal and NC terminal should indicate continuity with the door opened.

5. If improper indications are given, replace the defective switch.

6. Re-connect the leads to the interlock monitor switch terminals.

LIGHT CONTROL SWITCH TEST

The light control switch mounted on a bracket near the cavity rollers controls the cavity light operation. The switch operation can be checked with an ohmmeter.

1. Unplug the oven power cord, remove the outer case and discharge the capacitor.

2. Disconnect the harness connector from the light control switch.

3. With the door closed, a continuity check between the switch terminals COM (Common) and NO (Normally Open) should indicate continuity. A continuity check between terminals marked COM (Common) and NC (Normally closed) should indicate an open circuit.

4. Open the oven door. A continuity check between switch terminals marked COM and NO should now indicate an open circuit, while a continuity check between terminals marked COM and NC should indicate continuity with the oven door opened.

5. If improper indications are given, replace the defective switch and reconnect harness leads.

TIMER ASSEMBLY TEST

The Model 416.000 has a digital timer. If the timer motor does not operate with 120 volts a-c applied to the timer motor leads, the timer should be replaced.

The timer switch can be checked with an ohmmeter.

1. Unplug the oven power cord, remove the outer case and discharge the capacitor.

2. Disconnect the wire leads from the timer switch terminals.

3. Check continuity between the timer switch terminals marked COM and NO.

   a. With the timer in the OFF position, the ohmmeter should indicate an open circuit.

   b. With the timer turned to a selected time setting, the ohmmeter should indicate continuity.

SECTION 8 LITTON MICROWAVE OVENS

SERVICE PROCEDURE COMPONENT DATA

4. Replace the timer if defective, and re-install the harness leads.

SECTION G: COMPONENT REPLACEMENT AND ADJUSTMENT PROCEDURES

OUTER CASE REMOVAL

1. Turn the timer dial to OFF and disconnect the power cord.

2. Remove the mounting screws located on the back and sides of the outer case.

3. Raise the rear of the outer case slightly and slide the outer case back approximately one (1) inch.

4. Lift the entire case from the oven.

5. Discharge the capacitor by shorting the two capacitor terminals together with an insulated screwdriver.

NOTE: *When re-installing the outer case to the oven, be careful to properly align the exhaust duct gasket over the cavity exhaust duct.*

DOOR ASSEMBLY REPLACEMENT AND ADJUSTMENT

The following door replacement and adjustment procedures will normally limit R.F. leakage to less than 1 mw/cm^2 at 5 cm. Although maximum allowable leakage is 5 mw/cm^2 at 5 cm., effort should be made to assure leakage levels are well below the 5 mw/cm^2 maximum at 5 cm.

1. Unplug the oven power cord, remove outer case and discharge the capacitor.

2. Turn the oven on its back so that the oven door is facing upward.

3. Loosen the upper hinge mounting nuts.

4. Remove the door assembly from the oven.

5. Align the new door assembly to the oven front by centering the door hook alignment pin between the cavity rollers and sides of the access hole.

NOTE: *Be sure to install the nylon hinge spacer to the top and bottom hinge.*

6. Loosen the lower hinge mounting nuts.

7. Press the door against the cavity faceplate near the hinges and retighten the mounting nuts.

8. Re-adjust the door interlock switches as described under the Door Interlock Switch Adjustment in this section.

9. Check the oven for proper operation.

   Check the oven door operation for R.F. leakage around the door with an approved R.F. measuring device to assure less than 5 mw/cm^2 emission at 5 cm. in compliance with U.S. Government Department of Health, Education and Welfare 21 CFR, Subchapter J, Performance Standard for Microwave Ovens.

10. Re-install the outer wrapper.

DOOR INTERLOCK SWITCH ADJUSTMENT

1. Unplug the oven power cord, remove outer case and discharge the capacitor.

2. Loosen the three strike mounting screws that mount the strike assembly to the latch slide bracket.

3. With the oven door closed, align the strike assembly to the door hook to provide maximum activation of the door interlock switch.

4. Re-tighten the strike mounting screws.

5. Check the oven for proper door closure and switch operation.

6. Re-install the outer wrapper and check the oven for proper operation. Check the oven door operation for R.F. leakage around the door as outlined in Step 9 of the Door Replacement and Adjustment procedure.

CONTROL PANEL ASSEMBLY REMOVAL

1. Unplug the oven power cord, remove the outer case, and discharge the capacitor.

SECTION 8 LITTON MICROWAVE OVENS

SERVICE PROCEDURE COMPONENT DATA

2. Pull the door release handle from the latch slide bracket.

NOTE: *It may be necessary to force the handle off from the rear with a screwdriver, etc. When re-installing the handle, it may be necessary to install a new retaining spring.*

3. Remove solid-state controller knob.
4. Remove the five (5) mounting screws from the control panel assembly.
5. Remove the control panel assembly.

CAVITY LIGHT BULB REMOVAL (M)

1. Unplug the oven.
2. Remove the lamp cover mounting screw.
3. Remove the lamp cover.
4. Remove the cavity lamp.

COOK LIGHT REMOVAL (N)

1. Unplug the oven power cord, remove the outer case and discharge the capacitor.
2. Disconnect wire leads from the cook light.
3. Remove the control panel assembly as described in Control Panel Removal.
4. Remove the cook light by pushing light outward.

STIRRER COVER REMOVAL

1. Remove the plastic retaining rivets around the outer edges of the stirrer cover.
2. Remove the stirrer cover.

STIRRER MOTOR REMOVAL (P)

1. Unplug the oven power cord, remove the outer case and discharge the capacitor.
2. Remove the stirrer cover.

3. Remove the tinnerman clip from the stirrer shaft and pull down on the bottom spacer that is mounted directly under the stirrer blades.

NOTE: *The tinnerman clip and spacer can be turned in a downward direction for easier removal.*

4. Remove the stirrer blade assembly, being careful not to bend the blade assembly as it is pulled down from the shaft.
5. Remove the top spacer from the stirrer shaft.
6. Disconnect the stirrer leads from the wire harness.
7. Remove the two mounting screws and mounting nuts that hold the stirrer motor to the top of the cavity feedbox.
8. Remove the stirrer motor.

NOTE: *When re-installing the stirrer blade assembly to the stirrer shaft, observe the following procedures:*

1. Install the top spacer on the shaft with the smaller side up.
2. Install the stirrer blade assembly.
3. Install the bottom spacer with the smaller side up.
4. Install the tinnerman clip onto the stirrer shaft.
5. Turn the stirrer blades by hand to see that the blade assembly turns easily between the spacers.
6. Install the remaining components.

MAGNETRON ASSEMBLY REMOVAL (A)

1. Unplug the power cord, remove the outer case and discharge the capacitor.
2. Disconnect the wire leads from the magnetron filament terminals.
3. Remove the magnetron air ducts.
4. Remove the blower fan by pulling the fan from the motor shaft.
5. While supporting the magnetron, remove the four mounting nuts and washers.

SECTION 8 LITTON MICROWAVE OVENS

SERVICE PROCEDURE COMPONENT DATA

6. Carefully lower the magnetron, being careful not to damage the glass dome surrounding the antenna.

BLOWER MOTOR REMOVAL

1. Unplug the power cord, remove the outer case and discharge the capacitor.

2. Disconnect spade connectors from the blower motor terminals.

3. Remove the blower fan by pulling the fan from the motor shaft.

4. Remove the two blower motor mounting nuts and remove the blower motor.

5. The blower mounting brackets can be removed by removing the two mounting screws and nuts that mount the brackets to the oven base.

DIODE REMOVAL

1. Unplug the power cord, remove the outer case and discharge the capacitor.

2. Disconnect the diode from the capacitor.

3. Remove the diode mounting screw and nut from the oven base.

4. Remove the diode.

NOTE: *When installing the diode, keep diode body and leads away from any metal parts.*

POWER TRANSFORMER REMOVAL

1. Unplug the power cord, remove the outer case and discharge the capacitor.

2. Disconnect the primary wire leads and high-voltage lead from the power transformer.

3. Remove the four transformer mounting screws.

4. Remove the power transformer.

FILAMENT TRANSFORMER REMOVAL

1. Unplug the power cord, remove the outer case and discharge the capacitor.

2. Disconnect the primary and secondary wire leads.

3. Remove the two capacitor/filament transformer bracket mounting screws.

4. Remove the filament transformer.

CAPACITOR REMOVAL

1. Unplug the power cord, remove the outer case and discharge the capacitor.

2. Disconnect the spade connectors from the capacitor terminals.

3. Loosen the capacitor/filament transformer bracket mounting hardware.

4. Remove the capacitor.

DIGITAL TIMER ASSEMBLY REMOVAL

1. Unplug the power cord, remove the outer case and discharge the capacitor.

2. Loosen five control panel mounting screws.

3. Disconnect harness leads from the timer assembly.

4. Remove timer knob.

5. Remove three timer mounting screws.

6. Tilt control panel forward about one inch, tilt timer assembly sideways and slide backward to remove.

SHELF REPLACEMENT

1. Carefully remove the broken pieces of the shelf.

2. Remove any sealant that remains around the edges of the inner cavity.

3. Place a generous bead of sealant (approximately 1/4 inch) around the perimeter of the inner cavity edges to assure proper sealing of the new shelf.

4. Lay the shelft on the sealant and apply sufficient pressure to force out any excess sealant.

5. Wipe off the excess sealant and allow to cure for approximately twenty-four hours before using the oven.

CHOKE FILLER REPLACEMENT

1. Unplug the power cord, remove the outer case and discharge the capacitor.

2. Cut or break off the plastic mounting studs (located behind the cavity front) that hold the choke filler in place.

3. Insert the new choke filler into the cavity front.

4. Melt the plastic mounting studs with a soldering iron to retain the choke filler in place.

SOLID STATE CONTROLLER REMOVAL

1. Unplug the power cord, remove the outer case and discharge the capacitor.

2. Remove the control panel as described in the Control Panel Removal.

3. Remove vari-control pal-nut.

4. Disconnect wire leads from the controller.

5. Remove two controller mounting screws.

6. Remove controller from oven.

SECTION 8 LITTON MICROWAVE OVENS

SERVICE PROCEDURE COMPONENT DATA

SECTION H: WIRING SCHEMATIC, MODEL 416.000

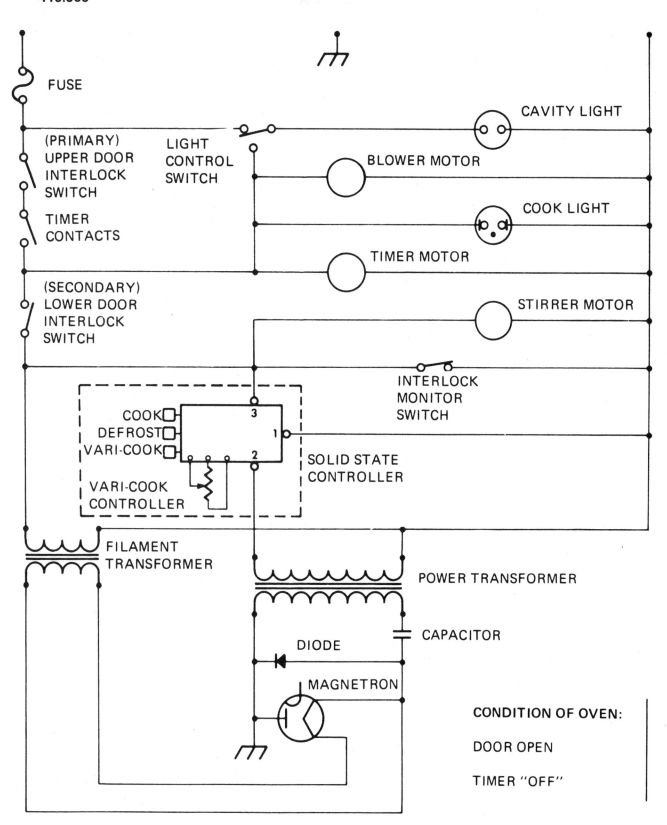

137

NOTES

GEM PRODUCTS, INC. MASTER PUBLICATIONS

APPLIANCE REPAIR MANUALS

REPAIR-MASTER® For Automatic Washers, Dryers & Dishwashers

These Repair-Masters® offer a quick and handy reference for the diagnosis and correction of service problems encountered on home appliances. They contain many representative illustrations, diagrams and photographs to clearly show the various components and their service procedure.

Diagnosis and repair charts provide step-by-step detailed procedures and instruction to solve the most intricate problems encountered in the repair of washers, dryers and dishwashers.

These problems range from timer calibrations to complete transmission repair, all of which are defined and explained in easy to read terms.

The Repair-Masters® are continually updated to note the latest changes or modifications in design or original parts. These changes and modifications are explained in their service context to keep the serviceman abreast of the latest developments in the industry.

AUTOMATIC WASHERS
- 9001 Whirlpool
- 9003 General Electric
- 9005 Westinghouse Front Loading
- 9009 Frigidaire
- 9010 Maytag
- 9011 Philco-Bendix Front Loading
- 9012 Speed Queen
- 9015 Norge – Plus Capacity
- 9016 Frigidaire Roller-Matic
- 9017 Westinghouse Top Loading

DISHWASHERS
- 5552 General Electric
- 5553 Kitchenaid
- 5554 Westinghouse
- 5555 D & M
- 5556 Whirlpool
- 5557 Frigidare

CLOTHES DRYERS
- 8051 Whirlpool-Kenmore
- 8052 General Electric
- 8053 Hamilton
- 8055 Maytag
- 8056 Westinghouse
- 8057 Speed Queen
- 8058 Franklin
- 8059 Frigidaire

*The D & M Dishwasher Repair-Masters covers the following brands: Admiral — Caloric — Chambers — Frigidaire (Portable) — Gaffers and Sattler — Gibson — Kelvinator — Kenmore — Magic Chef — Magic Maid — Norge — Philco — Pioneer — Preway — Roper — Wedgewood — Westinghouse (Portable)

The Repair-Masters® series will take the guesswork out of service repairs.

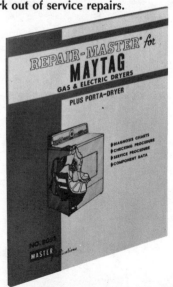

REPAIR-MASTER® for Domestic Refrigeration

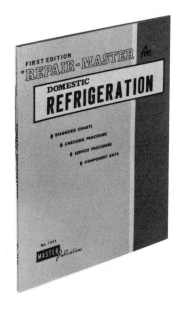

A BASIC GUIDE FOR REFRIGERATION REPAIR.

Takes the guesswork out of all your refrigeration repair work. The 7551 Repair-Master® covers the electrical, mechanical and hermetic systems of refrigeration. With the use of the diagnosis charts, malfunctions can be located and repaired rapidly.

- CHECKING PROCEDURE
- SERVICE PROCEDURE
- DIAGNOSIS CHARTS
- COMPONENT DATA

Valuable Information for the Refrigeration Servicetech!

7551
132 Pages

REPAIR-MASTER® for Domestic/Commercial Refrigeration

A most comprehensive Repair-Master® for the diagnosis and repair of both domestic and commercial refrigeration.

Includes electric testing, brazing, evacuation and recharging the refrigeration system. Step-by-step procedures for replacing components. Covers capillary tubes and expansion valves. Includes principles, cycle and the history of mechanical refrigeration.

This Repair-Master® is written in a manner that can be easily understood by the beginner and also serves as a complete refresher for the journeyman. A most comprehensive Repair-Master® that can be used for teaching refrigeration service.

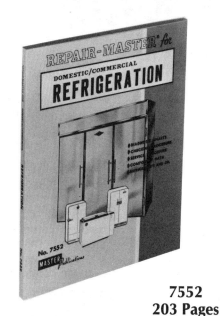

7552
203 Pages

- DIAGNOSIS CHARTS
- USE OF REFRIGERATION TOOLS
- HISTORY OF MECHANICAL REFRIGERATION
- PRINCIPLES OF REFRIGERATION
- THE REFRIGERATION CYCLE

- REFRIGERATION BY HEAT
- THE ABSORPTION SYSTEM
- THE GAS REFRIGERATOR
- COMPONENT REPLACEMENT
  AND MUCH MORE IS INCLUDED IN THIS REPAIR MASTER

REPAIR-MASTER® For Window Air Conditioners

7541
84 Pages

This Repair-Master® offers a quick and handy reference for the diagnosis and correction of service problems. Many illustrations, diagrams and photographs are included to clearly show the various components and their service procedures.

A Complete Service and Repair Guide!

PRINCIPLES OF REFRIGERATION
- DIAGNOSIS CHARTS
- TEST CORD TESTING
- PROCEDURES
- STEP-BY-STEP REPAIR
- SHOP PROCEDURES

REPAIR MASTER® For Domestic Automatic Icemakers

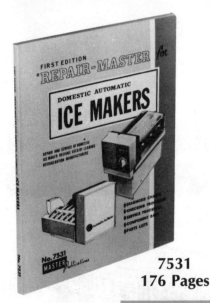

7531
176 Pages

Includes ten different designs of Automatic Ice Makers presently found in domestic refrigerators.

A special section is included for Whirlpool design self-contained Ice Maker. Illustrations, charts and new parts design information makes this Repair-Master® an important tool.

- DIAGNOSIS CHARTS
- TESTING PROCEDURES
- ADJUSTMENT PROCEDURES
- STEP-BY-STEP REPAIRS

TECH-MASTER® For Refrigerators & Freezers

For All Major Brands

Offering a quick reference by brand, year, and model number.

Handy 8½" x 5½" book is designed to be easily carried in the tool box or service truck. The first time you use these books they will more than repay the nominal purchase price.

Operating Information for Over 15,000 Models
- TEMPERATURE CONTROL PART NUMBERS
- CUT-IN AND CUT-OUT SETTINGS
- UNIT HORSEPOWER
- REFRIGERANT AND OIL CHARGE IN OUNCES
- SUCTION AND HEAD PRESSURES
- START AND RUN WATTAGES
- COMPRESSOR TERMINAL HOOK-UPS
- DEFROST HEATER WATTAGES

No. 7550	No. 7550-2	No. 7550-3	No. 7550-4
1950-1965	1966-1970	1971-1975	1976-1980
429 Pages	377 Pages	332 Pages	378 Pages